Project-Based Inquiry Science

ANIMALS IN ACTION

Janet L. Kolodner

Joseph S. Krajcik

Daniel C. Edelson

Brian J. Reiser

HERFF JONES EDUCATION DIVISION

IT's ABOUT TIME®
HERFF JONES EDUCATION DIVISION

84 Business Park Drive, Armonk, NY 10504
Phone (914) 273-2233 Fax (914) 273-2227
www.its-about-time.com

Program Components

Student Edition	Durable Equipment Kit
Teacher's Planning Guide	Consumable Equipment Kit
Teacher's Resources Guide	Multimedia
	—Observing and Interpreting Animal Behavior DVD
	—Flower Card Set

This project was supported, in part, by the **National Science Foundation** under grant nos. 0137807, 0527341, 0639978.
Opinions expressed are those of the authors and not necessarily those of the National Science Foundation.

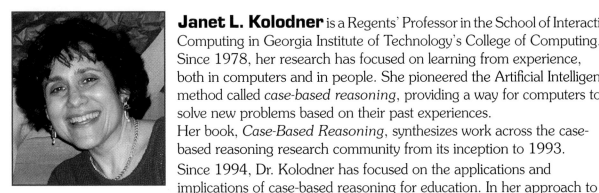

Principal Investigators

Janet L. Kolodner is a Regents' Professor in the School of Interactive Computing in Georgia Institute of Technology's College of Computing. Since 1978, her research has focused on learning from experience, both in computers and in people. She pioneered the Artificial Intelligence method called *case-based reasoning*, providing a way for computers to solve new problems based on their past experiences.

Her book, *Case-Based Reasoning*, synthesizes work across the case-based reasoning research community from its inception to 1993.

Since 1994, Dr. Kolodner has focused on the applications and implications of case-based reasoning for education. In her approach to science education, called Learning by Design™ (LBD), students learn science while pursuing design challenges. Dr. Kolodner has investigated how to create a culture of collaboration and rigorous science talk in classrooms, how to use a project challenge to promote focus on science content, and how students learn and develop when classrooms function as learning communities. Currently, Dr. Kolodner is investigating how to help young people come to think of themselves as scientific reasoners. Dr. Kolodner's research results have been widely published, including in *Cognitive Science, Design Studies,* and the *Journal of the Learning Sciences.*

Dr. Kolodner was founding Director of Georgia Techs' EduTech Institute, served as coordinator of Georgia Techs' Cognitive Science program for many years, and is founding Editor in Chief of the *Journal of the Learning Sciences.* She is a founder of the International Society for the Learning Sciences (ISLS), and she served as its first Executive Officer. She is a fellow of the American Association of Artificial Intelligence (AAAI).

Joseph S. Krajcik is a Professor of Science Education and Associate Dean for Research in the School of Education at the University of Michigan. He works with teachers in science classrooms to bring about sustained change by creating classroom environments in which students find solutions to important intellectual questions that subsume essential curriculum standards and use learning technologies as productivity tools. He seeks to discover what students learn in such environments, as well as to explore and find solutions to challenges that teachers face in enacting such complex instruction. Dr. Krajcik has authored and co-authored over 100 manuscripts and makes frequent presentations at international, national and regional conferences that focus on his research, as well as presentations that translate research findings into classroom practice. He is a fellow of the American Association for the Advancement of Science and served as president of the National Association for Research in Science Teaching. Dr. Krajcik co-directs the Center for Highly Interactive Classrooms, Curriculum and Computing in Education at the University of Michigan and is a co-principal investigator in the Center for Curriculum Materials in Science and The National Center for Learning and Teaching Nanoscale Science and Engineering. In 2002, Dr. Krajcik was honored to receive a Guest Professorship from Beijing Normal University in Beijing, China. In winter 2005, he was the Weston Visiting Professor of Science Education at the Weizmann Institute of Science in Rehovot, Israel.

Daniel C. Edelson is director of the Geographic Data in Education (GEODE) Initiative at Northwestern University where he is an Associate Professor of the Learning Sciences and Computer Science. Trained as a computer and cognitive scientist, Dr. Edelson develops and studies software and curricula that are informed by contemporary research on learning and motivation. Since 1992, Dr. Edelson has directed a series of projects exploring the use of technology as a catalyst for reform in science education and has led the development of a number of software environments for education. These include My World GIS, a geographic information system for inquiry-based learning, and WorldWatcher, a data visualization and analysis system for gridded geographic data, both of which have been recognized by educators for their contributions to Earth science education. Dr. Edelson is the author of the high school environmental science text, *Investigations in Environmental Science: A Case-Based Approach to the Study of Environmental Systems.* Dr. Edelson is currently engaged in research on professional development and implementation support for schools that have adopted *Investigations in Environmental Science.*

Since 1995, he has been the principal investigator on more than a dozen NSF-funded educational research and development grants, and he is a member of the leadership team of the NSF-funded Center for Curriculum Materials in Science. His research has been widely published, including in the *Journal of the Learning Sciences,* the *Journal of Research on Science Teaching,* the *Journal of Geoscience Education*, and *Science Teacher*.

Brian J. Reiser is a Professor of Learning Sciences in the School of Education and Social Policy at Northwestern University. Professor Reiser served as chair of Northwestern's Learning Sciences Ph.D. program from 1993, shortly after its inception, until 2001. His research focuses on the design and enactment of learning environments that support students' inquiry in science, including both science curriculum materials and scaffolded software tools. His research investigates the design of learning environments that scaffold scientific practices, including investigation, argumentation, and explanation; design principles for technology-infused curricula that engage students in inquiry projects; and the teaching practices that support student inquiry.

Professor Reiser also directed BGuILE (Biology Guided Inquiry Learning Environments) to develop software tools for supporting middle school and high school students in analyzing data and constructing explanations with biological data. Professor Reiser is a co-principal investigator in the NSF Center for Curriculum Materials in Science. He recently served as a member of the NRC panel authoring the report *Taking Science to School.* Professor Reiser received his Ph.D. in Cognitive Science from Yale University in 1983.

Acknowledgements

Three research teams contributed to the development of Project-Based Inquiry Science (PBIS): a team at the Georgia Institute of Technology headed by Janet L. Kolodner, a team at Northwestern University headed by Daniel Edelson and Brian Reiser, and a team at the University of Michigan headed by Joseph Krajcik and Ron Marx. Each of the PBIS units was originally developed by one of these teams and then later revised and edited to be a part of the full three-year middle-school curriculum that became PBIS.

PBIS has its roots in two educational approaches, Project-Based Science and Learning by Design™. Project-Based Science suggests that students should learn science through engaging in the same kinds of inquiry practices scientists use, in the context of scientific problems relevant to their lives and using tools authentic to science. Project-Based Science was originally conceived in the hi-ce Center at University of Michigan, with funding from the National Science Foundation. Learning by Design™ derives from Problem-Based Learning and suggests sequencing, social practices, and reflective activities for promoting learning. It engages students in design practices, including the use of iteration and deliberate reflection. LBD was conceived at Georgia Institute of Technology, with funding from the National Science Foundation, DARPA, and the McDonnell Foundation.

The development of the integrated PBIS curriculum was supported by the National Science Foundation under grants nos. 0137807, 0527341, and 0639978. Any opinions, findings and conclusions, or recommendations expressed in this material are those of the authors and do not necessarily reflect the views of the National Science Foundation.

PBIS Team

Principal Investigator
Janet L. Kolodner

Co-Principal Investigators
Daniel C. Edelson
Joseph S. Krajcik
Brian J. Reiser

NSF Program Officer
Gerhard Salinger

Curriculum Developers
Michael T. Ryan
Mary L. Starr

Teacher's Edition Developers
Rebecca M. Schneider
Mary L. Starr

Literacy Specialist
LeeAnn M. Sutherland

NSF Program Reviewer
Arthur Eisenkraft

Project Coordinator
Juliana Lancaster

External Evaluators
The Learning Partnership
Steven M. McGee
Jennifer Witers

The Georgia Institute of Technology Team

Project Director:
Janet L. Kolodner

Development of PBIS units at the Georgia Institute of Technology was conducted in conjunction with the Learning by Design™ Research group (LBD), Janet L. Kolodner, PI.

Lead Developers, Physical Science:
David Crismond
Michael T. Ryan

Lead Developer, Earth Science:
Paul J. Camp

Assessment and Evaluation:
Barbara Fasse
Jackie Gray
Daniel Hickey
Jennifer Holbrook
Laura Vandewiele

Project Pioneers:
JoAnne Collins
David Crismond
Joanna Fox
Alice Gertzman
Mark Guzdial
Cindy Hmelo-Silver
Douglas Holton
Roland Hubscher
N. Hari Narayanan
Wendy Newstetter
Valery Petrushin
Kathy Politis
Sadhana Puntambekar
David Rector
Janice Young

The Northwestern University Team

Project Directors:
Daniel Edelson
Brian Reiser

Lead Developer, Biology:
David Kanter

Lead Developers, Earth Science:
Jennifer Mundt Leimberer
Darlene Slusher

Development of PBIS units at Northwestern was conducted in conjunction with:

The Center for Learning Technologies in Urban Schools (LeTUS) at Northwestern, and the Chicago Public Schools
Clifton Burgess, PI
for Chicago Public Schools;
Louis Gomez, PI.

The BioQ Collaborative
David Kanter, PI.

The Biology Guided Learning Environments (BGuILE) Project
Brian Reiser, PI.

The Geographic Data in Education (GEODE) Initiative
Daniel Edelson, Director

The Center for Curriculum Materials in Science at Northwestern
Daniel Edelson,
Brian Reiser,
Bruce Sherin, PIs.

The University of Michigan Team

Project Directors:
Joseph Krajcik
Ron Marx

Literacy Specialist:
LeeAnn M. Sutherland

Project Coordinator:
Mary L. Starr

Development of PBIS units at the University of Michigan was conducted in conjunction with:

The Center for Learning Technologies in Urban Schools (LeTUS)
Phyllis Blumenfeld,
Barry Fishman,
Joseph Krajcik,
Ron Marx,
Elliot Soloway, PIs.

The Detroit Public Schools
Juanita Clay-Chambers
Deborah Peek-Brown

The Center for Highly Interactive Computing in Education (hi-ce)
Phyllis Blumenfeld,
Barry Fishman,
Joseph Krajcik,
Ron Marx,
Elizabeth Moje,
Elliot Soloway,
LeeAnn Sutherland, PIs.

Field-Test Teachers

National Field Test
Tamica Andrew
Leslie Baker
Jeanne Bayer
Gretchen Bryant
Boris Consuegra
Daun D'Aversa
Candi DiMauro
Kristie L. Divinski
Donna M. Dowd
Jason Fiorito
Lara Fish
Christine Gleason
Christine Hallerman
Terri L. Hart-Parker
Jennifer Hunn
Rhonda K. Hunter
Jessica Jones
Dawn Kuppersmith
Anthony F. Lawrence
Ann Novak
Rise Orsini
Tracy E. Parham
Cheryl Sgro-Ellis
Debra Tenenbaum
Sarah B. Topper
Becky Watts
Debra A. Williams
Ingrid M. Woolfolk
Ping-Jade Yang

New York City Field Test
Several sequences of PBIS units have been field-tested in New York City under the leadership of Whitney Lukens, Staff Developer for Region 9, and Greg Borman, Science Instructional Specialist, New York City Department of Education

6th Grade
Norman Agard
Tazinmudin Ali
Heather
   Guthartz Aniba
Asher Arzonane
Asli Aydin

Joshua Blum
Filomena Borrero
Shareese Blakely
John J. Blaylock
Tsedey Bogale
Zachary Brachio
Thelma Brown
Alicia Browne-Jones
Scott Bullis
Maximo Cabral
Lionel Callender
Matthew Carpenter
Ana Maria Castro
Diane Castro
Anne Chan
Ligia Chiorean
Boris Consuegra
Careen Halton Cooper
Cinnamon Czarnecki
Kristin Decker
Nancy Dejean
Gina DiCicco
Donna Dowd
Lizanne Espina
Joan Ferrato
Matt Finnerty
Jacqueline Flicker
Helen Fludd
Leigh Summers Frey
Helene Friedman-Hager
Diana Gering
Matthew Giles
Lucy Gill
Steven Gladden
Greg Grambo
Carrie Grodin-Vehling
Stephan Joanides
Kathryn Kadei
Paraskevi Karangunis
Cynthia Kerns
Martine Lalanne
Erin Lalor
Jennifer Lerman
Sara Lugert
Whitney Lukens
Dana Martorella
Christine Mazurek
Janine McGeown
Chevelle McKeever
Kevin Meyer
Jennifer Miller

Nicholas Miller
Diana Neligan
Caitlin Van Ness
Marlyn Orque
Eloisa Gelo Ortiz
Gina Papadopoulos
Tim Perez
Albertha Petrochilos
Christopher Poli
Kristina Rodriguez
Nadiesta Sanchez
Annette Schavez
Hilary Sedgwitch
Elissa Seto
Laura Shectman
Audrey Shmuel
Katherine Silva
Ragini Singhal
C. Nicole Smith
Gitangali Sohit
Justin Stein
Thomas Tapia
Eilish Walsh-Lennon
Lisa Wong
Brian Yanek
Cesar Yarleque
David Zaretsky
Colleen Zarinsky

7th Grade
Mayra Amaro
Emmanuel Anastasiou
Cheryl Barnhill
Bryce Cahn
Ligia Chiorean
Ben Colella
Boris Consuegra
Careen Halton Cooper
Elizabeth Derse
Urmilla Dhanraj
Gina DiCicco
Lydia Doubleday
Lizanne Espina
Matt Finnerty
Steven Gladden
Stephanie Goldberg
Nicholas Graham
Robert Hunter
Charlene Joseph
Ketlynne Joseph
Kimberly Kavazanjian

Christine Kennedy
Bakwah Kotung
Lisa Kraker
Anthony Lett
Herb Lippe
Jennifer Lopez
Jill Mastromarino
Kerry McKie
Christie Morgado
Patrick O'Connor
Agnes Ochiagha
Tim Perez
Nadia Piltser
Chris Poli
Carmelo Ruiz
Kim Sanders
Leslie Schiavone
Ileana Solla
Jacqueline Taylor
Purvi Vora
Ester Wiltz
Carla Yuille
Marcy Sexauer Zacchea
Lidan Zhou

ANIMALS IN ACTION

Animals in Action

The PBIS version of *Animals in Action* is an adaptation of the unit *Animals in Action* developed by a team at University of Toledo. The unit was inspired by and includes activities adapted from *Behavior Matters*, developed jointly by the Brookfield Zoo in Chicago and the Center for Learning Technologies in Urban Schools at Northwestern University.

Animals in Action

PBIS Editorial Team:

Francesca Casella
Mary L. Starr

University of Toledo Team

Lead Developer

Rebecca M. Schneider

Other Developers

Cara Aschliman
Jodie VonSeggern

Pilot Teachers

Lara Fish
Cheri Sgro-Ellis
Debra Tennenbaum

Behavior Matters

Project Directors

Brian J. Reiser
Susan Margulis

Writers/Editors

Marilyn Havlik
Debora Ward

Contributors and Pilot Teachers

Steve Arnam
Jerry Balin
Cindy Cho
Robin Dembeck
Michele Fleming
Vanessa Go
Ravit Golan
Carl Koch
Lanis Petrik
Brian Reiser
Keith Winsten
Angela Wrobel
Sherry Yarema

The development of *Animals in Action* was supported by the National Science Foundation under grants no. 0137807, 0527341, 0639978. The development of *Behavior Matters* was funded by the National Science Foundation under grant nos. 9809636 and 9720383. We are grateful for the recommendations of Whitney Lukens of the NYC Public Schools during development of the PBIS Unit. Any opinions, findings, and conclusions or recommendations expressed in this material are those of the authors and do not necessarily reflect the views of the National Science Foundation or the Brookfield Zoo.

Table of Contents

What's the Big Question?

How Do Scientists Answer Big Questions and Solve Big Problems? ....AIA 3

Learning Set 1

Ethology, instinctive and learned behavior, adaptation, species, survival, social interaction, collecting observational data, careful observation, keeping good records, finding trends in data, reliable data, interpretation, using evidence to support claims, explanation, collaboration, iteration, carnivores, herbivores, omnivores

How Do Biologists Study Animal Behavior? .....AIA 11

1.1 Understand the Question
Observing Animal Behavior... AIA 12

1.2 Explore
How Can You Improve Your Data Collection .... AIA 15

1.3 Explore
Observing and Interpreting Animal Behavior ....AIA 20

1.4 Explain
Support Your Interpretation .... AIA 28

1.5 Read
What Do Animals Need to Survive? .... AIA 32

Back to the Big Question ....AIA 38

Learning Set 2

Foragers, predators, structure and function, effects of habitat, adaptation, mutualistic relationships, light, color, animal vision, joints and levers, careful observation, keeping good records, finding trends in data, reliable data, interpretation, using evidence to support claims, explanation, collaboration, building on the work of others, models and simulations

What Affects How Animals Feed? ....AIA 43

2.1 Understand the Question
Thinking about What Affects How Animals Feed .... AIA 44

2.2 Explore
What Affects How Chimpanzees Feed? .... AIA 47

2.3 Read
How Do Chimpanzees Feed and Why?.......................... AIA 53

2.4 Investigate
How Do Bees Forage? .... AIA 61

2.5 Investigate
What Do Bees See that Helps Them Forage? .... AIA 68

More to Learn
Light and Color Vision... AIA 72

2.6 Read
What Adaptations Do Bees Have that Affect Their Feeding Behavior?............................. AIA 74

More to Learn
Flower Dissection.. AIA 78

2.7 Explore
What Are the Feeding Behaviors of Some Other Carnivores?... AIA 82

Back to the Big Challenge................................. AIA 91

Learning Set 3

Verbal and non-verbal communication, structure and function, effects of habitat, adaptation, the waggle dance, echolocation, sonar, sound and sound waves, how humans hear, careful observation, keeping good records, finding trends in data, reliable data, interpretation, using evidence to support claims, explanation, collaboration, building on the work of others, criteria and constraints

What Affects How Animals Communicate? .... AIA 97

3.1 Understand the Question
Thinking about What Affects How Animals Communicate .... AIA 98

3.2 Investigate
How Do Humans Communicate?............................... AIA 102

3.3 Explore
How Do Bees Communicate and Why? .... AIA 109

3.4 Explore
What Affects How Elephants Communicate? ....AIA 114

3.5 Read
How Are Elephants Adapted for Communication? .... AIA 120

More to Learn
What Is Sound?.. AIA 124

3.6 Explore
What Affects How Marine Mammals Communicate? ... AIA 131

3.7 Read
How Do Dolphins Communicate?.............................. AIA 136

Back to the Big Challenge .... AIA 142

Address the Big Challenge

Design an Enclosure for a Zoo Animal that Will Allow it to Feed or Communicate as in the Wild .... AIA 145

Answer the Big Question

How Do Scientists Answer Big Questions and Solve Big Problems .... AIA 154

Glossary, Index, and Picture Credits ....AIA 156

Welcome to Project-Based Inquiry Science!

Dear Students,

This year, you will be learning the way scientists learn. You will explore interesting questions and challenges. You will learn new things. You will also learn exciting, new ways to think about the world around you.

Scientists learn as they are trying to answer a big question or solve a big challenge. To help them work on these big questions or challenges, scientists break them into smaller ones. For each smaller question or challenge, they read what other scientists already know, and they investigate, explore, gather evidence, and form explanations. This way, scientists build new knowledge as they answer these smaller questions. Then they use what they have learned to try to answer the big question or solve the big challenge. Along the way, scientists share what they have learned with other scientists. These other scientists can then use this new knowledge to address other questions and challenges.

Like scientists, you will be trying to answer big questions and solve big challenges this year. You will break these into smaller questions or challenges. For each smaller question or challenge, you'll read, investigate, explore, gather evidence, and form explanations. As you do these things, you will share what you are learning and work closely with your classmates. You and your classmates will help each other learn and successfully answer each unit's question or solve its challenge. At the end of each unit, you'll answer the big question or address the big challenge based on what you've learned. And you will have learned a lot!

PBIS was designed to support you as you become a student scientist. In fact, PBIS units were written by scientists who study how people learn and who want to help you become the best scientist you can be. We used what we know about learning to design ways to help you answer big questions and solve big challenges. What you learn this year about science will help you learn science in the future. The way you learn to think about questions and challenges will help you learn other subjects, too.

Each year begins with a launcher unit. Launcher units help your class learn to work together, help you become familiar with the ways scientists think and have discussions, and introduce you to the activities and tools you'll use throughout PBIS.

Have fun being a student scientist!

Janet L. Kolodner

Janet Kolodner (for the whole PBIS team)

Introducing PBIS

What Do Scientists Do?

1) Scientists...address big challenges and big questions.

You will find many different kinds of *Big Challenges* and *Questions* in PBIS Units. Some ask you to think about why something is a certain way. Some ask you to think about what causes something to change. Some challenge you to design a solution to a problem. Most of them are about things that can and do happen in the real world.

Understand the Big Challenge or Question

As you get started with each Unit, you will do activities that help you understand the *Big Question* or *Challenge* for that Unit. You will think about what you already know that might help you, and you will identify some of the new things you will need to learn.

Project Board

The *Project Board* helps you keep track of your learning. For each challenge or question, you will use a *Project Board* to keep track of what you know, what you need to learn, and what you are learning. As you learn and gather evidence, you will record that on the *Project Board*. After you have answered each small question or challenge, you will return to the *Project Board* to record how what you've learned helps you answer the *Big Question* or *Challenge*.

Learning Sets

Each Unit is composed of a group of *Learning Sets*, one for each of the smaller questions that needs to be answered to address the *Big Question* or *Challenge*. In each *Learning Set*, you will investigate and read to find answers to the *Learning Set's* question. You will also have a chance to share the results of your investigations with your classmates and work together to make sense of what you are learning. As you come to understand answers to the questions on the *Project Board*, you will record those answers and the evidence you've collected that convinces you of what you've learned. At the end of each *Learning Set*, you will apply what you've learned to the *Big Question* or *Challenge*.

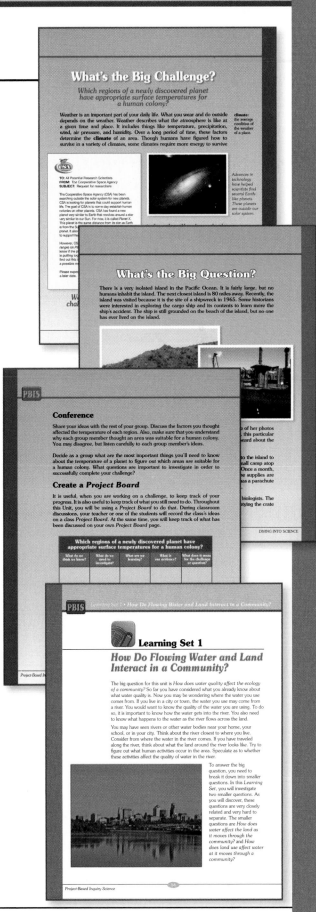

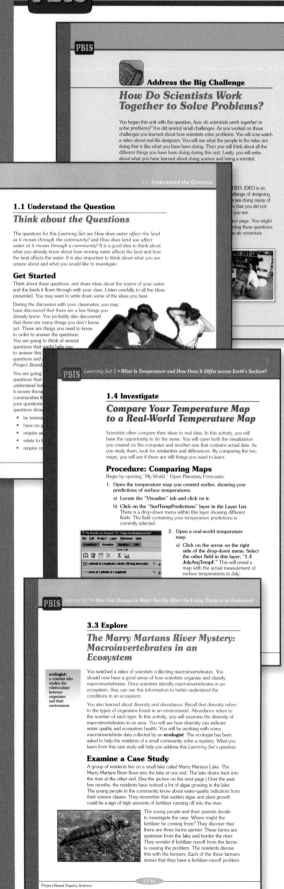

Address the Big Challenge/ Answer the Big Question

At the end of each Unit, you will put everything you have learned together to tackle the *Big Challenge* or *Question*.

2) Scientists...address smaller questions and challenges.

What You Do in a Learning Set

Understanding the Question or Challenge

At the start of each *Learning Set*, you will usually do activities that will help you understand the *Learning Set's* question or challenge and recognize what you already know that can help you answer the question or achieve the challenge. Usually, you will visit the *Project Board* after these activities and record on it the even smaller questions that you need to investigate to answer a *Learning Set's* question.

Investigate/Explore

There are many different kinds of investigations you might do to find answers to questions. In the *Learning Sets* you might

- design and run experiments;
- design and run simulations;
- design and build models
- examine large sets of data.

Don't worry if you haven't done these things before. The text will provide you with lots of help in designing your investigations and in analyzing your data.

Read

Like scientists, you will also read about the science you are learning. You'll read a little bit before you investigate, but most of the reading you do will be to help you understand what you've experienced or seen in an investigation. Each time you read, the text will include *Stop and Think* questions after the reading. These questions will help you gauge how well you understand what you have read. Usually, the class will discuss the answers to *Stop and Think* questions before going on so that everybody has a chance to make sense of the reading.

Design and Build

When the *Big Challenge* for a Unit asks you to design something, the challenge in a *Learning Set* might also ask you to design something and make it work. Often, you will design a part of the thing you will design and build for the big challenge. When a *Learning Set* challenges you to design and build something, you will do several things:

- identify what questions you need to answer to be successful
- investigate to find answers to those questions
- use those answers to plan a good design solution
- build and test your design

Because designs don't always work the way you want them to, you will usually do a design challenge more than once. Each time through, you will test your design. If your design doesn't work as well as you'd like, you will determine why it is not working and identify other things you need to learn to make it work better. Then, you will learn those things and try again.

Explain and Recommend

A big part of what scientists do is explain, or try to make sense of why things happen the way they do. An explanation describes why something is the way it is or behaves the way it does. An explanation is a statement you make built from claims (what you think you know), evidence (from an investigation) that supports the claim, and science knowledge. As they learn, scientists get better at explaining. You'll see that you get better, too, as you work through the *Learning Sets*.

A recommendation is a special kind of claim—one where you advise somebody about what to do. You will make recommendations and support them with evidence, science knowledge, and explanations.

ANIMALS IN ACTION

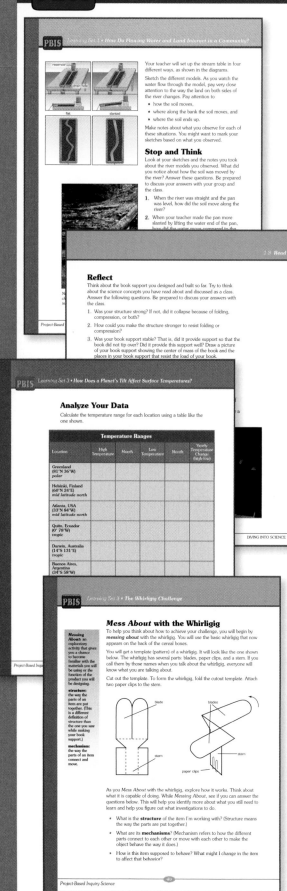

3) Scientists...reflect in many different ways.

PBIS provides guidance to help you think about what you are doing and to recognize what you are learning. Doing this often as you are working will help you be a successful student scientist.

Tools for Making Sense

Stop and Think

Stop and Think sections help you make sense of what you've been doing in the section you are working on. *Stop and Think* sections include a set of questions to help you understand what you've just read or done. Sometimes the questions will remind you of something you need to pay more attention to. Sometimes they will help you connect what you've just read to things you already know. When there is a *Stop and Think* in the text, you will work individually or with a partner to answer the questions, and then the whole class will discuss what you've learned.

Reflect

Reflect sections help you connect what you've just done with other things you've read or done earlier in the Unit (or in another Unit). When there is a *Reflect* in the text, you will work individually, with a partner or your small group to answer the questions. Then, the whole class will discuss what you've learned. You may be asked to answer *Reflect* questions for homework.

Analyze Your Data

Whenever you have to analyze data, the text will provide hints about how to do that and what to look for.

Mess About

"Messing about" is a term that comes from design. It means exploring the materials you will be using for designing or building something or examining something that works like what you will be designing. Messing about helps you discover new ideas—and it can be a lot of fun. The text will usually give you ideas about things to notice as you are messing about.

What's the Point?

At the end of each *Learning Set*, you will find a summary, called *What's the Point?*, of the important things we hope you learned from the *Learning Set*. These summaries can help you remember how what you did and learned is connected to the *Big Challenge* or *Question* you are working on.

4) Scientists...collaborate.

Scientists never do all their work alone. They work with other scientists (collaborate) and share their knowledge. PBIS helps you be a student scientist by giving you lots of opportunities for sharing your findings, ideas, and discoveries with others (the way scientists do). You will work together in small groups to investigate, design, explain, and do other things. Sometimes you will work in pairs to figure things out together. You will also have lots of opportunities to share your findings with the rest of your classmates and make sense together of what you are learning.

Investigation Expo

In an *Investigation Expo*, small groups report to the class about an investigation they've done. For each *Investigation Expo*, you will make a poster detailing what you were trying to learn from your investigation, what you did, your data, and your interpretation of your data. The text gives you hints about what to present and what to look for in other groups' presentations. *Investigation Expos* are always followed by discussions about what you've learned and about how to do science well. You may also be asked to write a lab report following an investigation.

Plan Briefing/Solution Briefing/Idea Briefing

Briefings are presentations of work in progress. They give you a chance to get advice from your classmates that can help you move forward. During a *Plan Briefing*, you present your plan to the class. It might be a plan for an experiment or a plan for solving a problem or achieving a challenge. During a *Solution Briefing*, you present your solution in progress and ask the class to help you make your solution better. During an *Idea Briefing*, you present your ideas. You get the best advice from your classmates when you present evidence in support of your plan, solution, or idea. Often, you will prepare a poster to help you make your presentation. Briefings are almost always followed by discussions of what you've learned and how you will move forward.

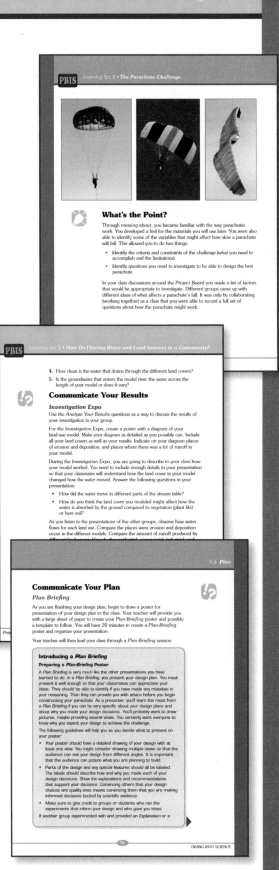

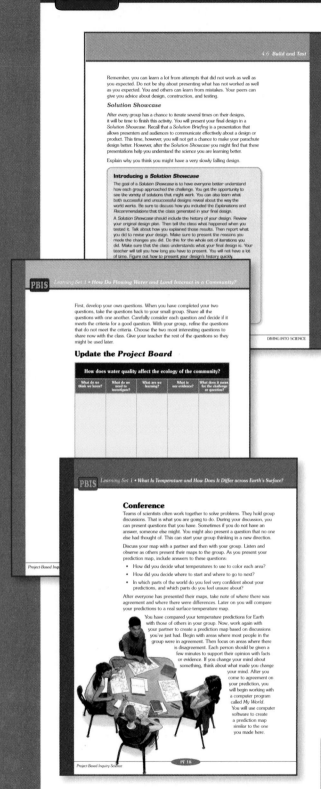

Solution Showcase

Solution Showcases usually appear near the end of a Unit. During a *Solution Showcase*, you show your classmates your finished product—either your answer to a question or your solution to a challenge. You also tell the class why you think it is a good answer or solution, what evidence and science you used to get to your solution, and what you tried along the way before getting to your answer or solution. Sometimes a *Solution Showcase* is followed by a competition. It is almost always followed by a discussion comparing and contrasting the different answers and solutions groups have come up with. You may be asked to write a report or paper following a *Solution Showcase*.

Update the *Project Board*

Remember that the *Project Board* is designed to help the class keep track of what they are learning and their progress towards a Unit's *Big Question* or *Challenge*. At the beginning of each Unit, the class creates a *Project Board*, and together you record what you think you know about answering the *Big Question* or addressing the *Challenge* and what you think you need to investigate further. Near the beginning of each *Learning Set*, the class revisits the *Project Board* and adds new questions and things they think they know. At the end of each *Learning Set*, the class again revisits the *Project Board*. This time you record what you have learned, the evidence you've collected, and recommendations you can make about answering the *Big Question* or achieving the *Big Challenge*.

Conference

A *Conference* is a short discussion between a small group of students before a more formal whole-class discussion. Students might discuss predictions and observations, they might try to explain together, they might consult on what they think they know, and so on. Usually, a *Conference* is followed by a discussion around the *Project Board*. In these small group discussions, everybody gets a chance to participate.

What's the Point?
Review what you have learned in each *Learning Set*.

Stop and Think
Answer questions that help you understand what you've done in a section.

Communicate
Share your ideas and results with your classmates.

Record
Record your data as you gather it.

ANIMALS IN ACTION

As a student scientist, you will...

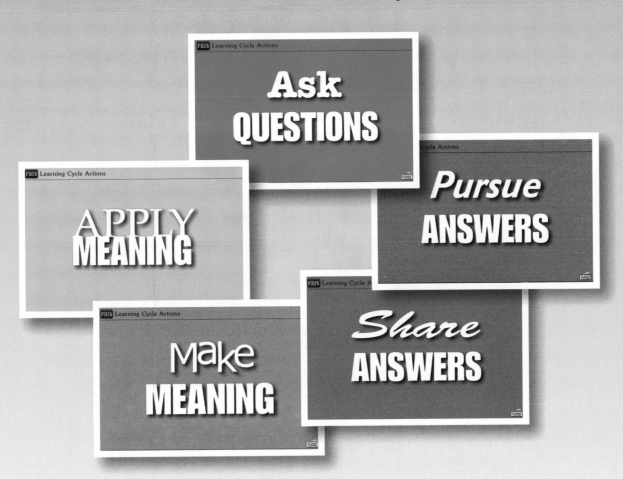

What's the Big Question?

How do scientists answer big questions and solve big problems?

Imagine that on your way to school one morning you see a bird. It swoops down to the ground then flies back into the tree. You stop to watch and wonder what the bird is doing. Why is it going back and forth between the ground and the tree? You see a small pile of crumbs on the ground. Then, you notice a nest built in the tree branches. All these things help you realize that the bird is feeding its young in the nest. When you take the time to watch the world in this way, you are acting much like **biologists** do.

biologist: a scientist who studies living things.

observe: to use one of the five senses to gather information about an object or phenomenon.

Biologists are a group of scientists who study living things. One thing they study is the behavior of animals. Studying animal behavior helps them better understand the living world. The results of their studies can help you to see animals in a new and different way. Through the study of animals, you may even better understand your own behavior or the behavior of your friends.

In this Unit, you will **observe** and study the behavior of several animals. You will learn about how animals feed and communicate with each other and what conditions affect those things. You will also develop some tools to help you collect and organize your observations. These tools will help you analyze your data as you try to answer the *Big Question: How do scientists answer big questions and solve big problems?* You will answer this question in the context of a science question: *Why do animals behave the way they do?*

Welcome to Animals In Action!
Enjoy being a student scientist.

Think about the Big Question

In this Unit, you will respond to a challenge to answer the *Big Question*. Before you start to think about what you already know about the *Big Question*, read about the challenge you will address.

Your Challenge

Look at the pictures on this page. They show animal **enclosures** found in zoos early in the 20th century. Some of these zoos were built over 100 years ago. Observe the pictures closely. The animals in these enclosures are all in cages. The zoos kept very large animals and smaller mammals, birds, and reptiles in similar enclosures. Zoos built these kinds of enclosures for animals at a time when zoos were designed for the display of animals and as places where people could have fun. In addition to animals, the zoo might also have had an amusement park, a playground, or a dance hall. The animal cages were usually very primitive. They were made of steel with cement floors and only sometimes included trees or water.

Zoos do not have amusement–park rides anymore. Today, zoos are concerned with **conservation** and education rather than the display of unusual animals. As the purposes of zoos have changed, so have spaces built for the animals. Nowadays, many zoos build animal spaces in ways that allow the animals to live more like they do in their natural habitat and allow people to learn about animals by watching them. For example, in the 1920s, the Detroit Zoo built a new home for birds. The large domed building held many cages for the birds to live in. The building was a bright and comfortable place for zoo visitors to watch the birds.

Toucan at the Central Park Zoo in New York City.

The Home of the Birds

Visitors to "The Zoo" invariably begin their tour of the park by visiting the bird house, located near the entrance gates. This enormous bird sanctuary was completed in 1927.

Of exquisite architecture, this handsome building has a 75-foot glass dome rising from a beautiful setting of shady maples and shrubbery. It is equipped with a total of 97 cages in which nearly 500 birds make their homes.

Cages for birds from tropical countries are constructed on the interior of the building, and on the western outside exposure there are cages for birds native to North America and the temperate zones.

Among these are pigeons, hawks, ruffly little owls and blackbirds. Here too are the Mandrill baboon "Betsey," "Snooty" the coatimundi captured by a U. S. Marine in Nicaragua, who though he hails from the tropics must have constant access to fresh air, and "Smitty" and "Jiggs," a species of monkey known as the drill.

The monkey cages are so arranged that their tenants may take refuge within the building or escape to fresh air and a change of scene on the outside of the building, at will.

A Big Bird Cage

Within the bird house, an almost tropical heat is maintained for the comfort of the birds collected from all over the world.

Sunshine floods down from the skylight on the giant cage in the center of the building. Within this cage, reside the glorious white egret, the ibis, gray and white, in sharp contrast to the jet black crows whose coarse cawing mingles with the trills and songs of more delicate and graceful birds.

In the cages at the sides are dainty love birds, the colorful toucans, the black and yellow tropiole, blue and yellow macaws, tree ducks, starling, finches, parrakeets, cockatoos, the broad-tail wydah, rose cockatoos, the orange weaver.

Europe, Asia, Africa and both of the Americas have contributed their feathered citizens of the air to the inhabitants of the bird house at "The Zoo." Each cage is labeled with the name and habitat of its residents. Not all the birds here are rare, but all are remarkable for color, grace or song. Each variety, too, has its own peculiarity of diet. The toucan, for example, must have its food rolled into pellets which it swallows as one would a pill.

The Detroit Zoo's Bird exhibit, built in the 1920s.

This panda's exhibit at the National Zoo, in Washington, DC, is based on the needs of the panda.

In 1996, the Detroit Zoo renovated the old bird house transforming it into a new butterfly house and interpretive center. Now birds, as well as butterflies, are free to fly within the building. There are many plants for shelter and water in the renovated exhibit. The space created for the animals has changed, and so has what people can learn by watching the animals.

One goal of zoos is to make the zoo **habitat** as close to the animal's natural environment as possible. When animals live in areas that look more like their natural surroundings, they are more likely to act naturally. This way, biologists can find out more about animal behavior, and zoo visitors can better see how animals behave in their natural habitats.

Recently, the panda enclosure at the National Zoo in Washington, DC was updated. The National Zoo is committed to making the captive animals' lives as similar to their natural life as possible. The design of the new panda area was based on scientists' observations of pandas in the field. To determine what important features the new environment would need, scientists watched as the pandas ate, played, slept, and interacted with one another. The pandas are thriving in the new habitat, and zoo visitors and scientists are learning more about these animals.

enclosure: an area that is surrounded by something like a fence or a wall.

conservation: the preservation, management, and care of natural and cultural resources.

habitat: a place where animals (including people) live.

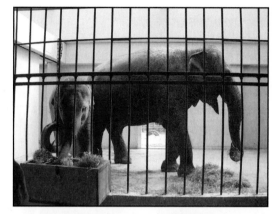

WHAT'S NEW IN ZOOS

All zoos are putting the needs of the animals first. In 2004, Michigan's Detroit Zoo was one of the first zoos to permanently close its elephant exhibit and retire two female Asian elephants, Winky and Wanda, to a sanctuary. The Detroit Zoo chose to do this for ethical reasons. Detroit Zoo Director Ron Kagan stated, "Now we understand how much more is needed to be able to meet all the physical and psychological needs of elephants in captivity, especially in a cold climate."

Winky, now age 51, and Wanda, now age 46, were captured in the wild as babies and have been companions at the Detroit Zoo since 1994. Thanks to the Detroit Zoo's humane decision, Winky and Wanda will enjoy full, enriching years of retirement, roaming through hundreds of acres of natural habitat in the company of many other elephants.

Your challenge for this Unit is to design a new enclosure that will accommodate the feeding or communication of one of the animals you study in this Unit. Your goal will be to design the zoo environment so it is similar enough to the natural environment of the animal to allow the animal to feed or communicate effectively. The enclosure will also have to allow visitors and scientists to observe the animals clearly.

Identify Criteria and Constraints

Before getting started on a challenge, it is important to make sure you understand the challenge. Design challenges have two parts: **criteria** and **constraints**.

Criteria are goals that must be satisfied to achieve the challenge. For the zoo enclosure challenge, this will include designing the enclosure so that the animals it will hold can communicate or feed as they would in their natural habitat. It will have to allow scientists and zoo visitors to observe the animals effectively.

Constraints are factors that limit how you can address a challenge. Your biggest constraint will be that the enclosure be built close to where you live. You will be able to assume that space can be found, but you will have to think about the weather where you live and how it will affect your enclosure design. You can probably think of other constraints.

With your class, identify the full set of criteria and constraints for this challenge, and put them on a chart like the one below so that you will remember them as you move through the Unit.

criteria: (singular, criterion) goals that must be satisfied to successfully achieve a challenge.

constraints: factors that limit how you can achieve a challenge.

The Zoo Enclosure Challenge	
Criteria	Constraints
The animal you choose has to be able to feed or communicate as it would in its natural habitat.	The weather where we live will require...
It has to be easy for zoo visitors to observe the feeding or communication.	

Create a *Project Board*

Project Board:
a chart for keeping track of progress as you work on a project over a long period of time.

In this Unit, you will be working toward achieving a *Big Challenge*. This Unit, like other Units in *Project-Based Inquiry Science (PBIS)*, is broken into *Learning Sets*. Each *Learning Set* helps you learn a different set of concepts and skills. At the end of each *Learning Set*, you will work toward applying what you have just learned to the *Big Challenge*. Then at the end of the whole Unit, you will return to the *Big Challenge* again to pull everything together and create a solution.

When you work on a big project, it is useful to keep track of your progress and what you still need to do. You will use a *Project Board* to do that.

Be a Scientist

Introducing the *Project Board*

When you work on a project, it is useful to keep track of your progress and what you still need to do. A *Project Board* gives you a place to keep track of your scientific understanding as you make your way through a Unit. It is designed to help your class organize its questions, investigations, results, and conclusions. The *Project Board* will also help you decide what to do next. During classroom discussions, you will record the class's ideas on a class *Project Board*. At the same time, you will also keep your own *Project Board* page.

The *Project Board* has space for answering five guiding questions:

- What do we think we know?
- What do we need to investigate?
- What are we learning?
- What is our evidence?
- What does it mean for the challenge or question?

Each time you use the *Project Board*, you will record as much as you can in each column. As you work through a Unit, you will return over and over again to the *Project Board*. You will add more information and revise what you have recorded. Everything you write in the columns will be based on what you know or what you have learned. In addition to text, you will sometimes want to put pictures or data on the board.

To get started on this *Project Board*, review the questions you are answering and the challenge you are addressing. Your challenge is to design an enclosure for an animal that will allow the animal to behave as it would in its natural habitat and that will allow visitors and scientists to observe and study the animal. This will help you answer the questions: *How do scientists answer big questions and solve big problems?* and *Why do animals behave the way they do?* Record these questions in the top area of the *Project Board* as shown below.

How do scientists answer big questions and solve big problems? Why do animals behave the way they do?				
What do we think we know?	What do we need to investigate?	What are we learning?	What is our evidence?	What does it mean for the challenge or question?

As you create your *Project Board* for this Unit, you will focus on animal behavior and on making good observations. Think about what you know that would help you address the challenge. You might have experiences with animal behavior, the jobs of biologists, or making observations and inferences that will be important for addressing this challenge. You may also have some questions about zoos, zoo animals, or studying animals.

You will begin by focusing on the first two columns: *What do we think we know?* and *What do we need to investigate?*

What do we think we know?

In this column, you will record what you think you know that is important to the challenge. This might be what you know about animal behavior or about studying animal behavior. You might also want to record what you know about animal enclosures or the behavior of animals in captivity. You probably think you know a lot about animal behavior. Some things may not be completely accurate. It is important to record those things anyway, for two reasons:

- When you look at the *Project Board* later, you will be able to see how much you have learned.

- Discussion with your class will help you figure out what you need to investigate.

What do we need to investigate?

In this column, you will record the things you need to learn more about to address the challenge. You probably have many ideas now about what you need to investigate. Work with your class to get these ideas on the *Project Board*. Later in this Unit, you will add other questions. Later, you may find things you are confused about. You and your classmates might disagree about some ideas. You will be recording in this column what you do not understand well or what you disagree about.

Sometimes you will be unsure about how to word your idea as a question. One of the things your class will do together around the *Project Board* is turn the things you are curious about into questions you can investigate.

You will return to the *Project Board* many times in this Unit. You will continue to add information to the board. You will record many of the ideas you have and things you are learning. You will then see how your ideas change. By the end of the Unit, you will fill in all of the columns.

The *Project Board* is a great place to start discussions. You may find that you disagree with other classmates about what you have learned and the evidence for it. This is a part of what scientists do. Such discussions help participants identify what they or others do not understand well and what else they need to learn or investigate. The class will fill in the large *Project Board*. Make sure to record the same information on your own *Project Board* page.

Learning Set 1

How Do Biologists Study Animal Behavior?

Animals are interesting to people. Humans work hard to understand why animals behave the way they do. People have been observing animals for hundreds of years. The *Big Question* for this Unit is: *How do scientists answer big questions and solve big problems?* You will answer this question in the context of the science question: *Why do animals behave the way they do?* You will apply what you are learning to designing a zoo enclosure. Before you start your observations, you need to break the *Big Question* into smaller questions. The smaller question you will answer in this *Learning Set* is: *How do biologists study animal behavior?* To answer this question, it will be important to work as biologists work, by making careful observations. You will be thinking about how to gather, record, and analyze data about animals in a way biologists do.

Notice how the animals in these pictures behave differently. The biologists who study animal behavior make careful observations and gather, record, and analyze data to determine why animals behave the way they do.

1.1 Understand the Question

Observing Animal Behavior

data: (singular, datum) recorded measurements or observations.

ethologist: a biologist who studies the behavior of animals in their natural environment.

Whether you are observing animals, people, machines, or stars, observation is a critical science skill. Without careful observations, scientists could not explain the way the world works. When scientists make observations, they gather a lot of **data**. The data serve as a record of what the scientists observed. By making a record, a scientist can return to the data often and for different purposes. Without a record, the scientist would just be working from memory, and memory is not always reliable.

Ethologists are biologists who study how animals behave. They observe animals to explain why animals behave the way they do in their natural environment. Ethologists usually observe animals to answer the following questions:

- How do different behaviors help an animal survive?

- How do the animal's environment and learning affect different behaviors?

- How do behaviors change as the animal grows?

- How do animals that are similar to each other act in similar or different ways?

To answer these four questions, ethologists observe very carefully and for long periods of time. Ethologists collect a lot of information, or data. To make sense of all the data they collect, ethologists use specific rules to describe and classify what they observe.

You, too, can be an ethologist as you investigate the behavior of animals. In this Unit, you will think about some of these questions. In this *Learning Set*, you will develop some of the tools and observation methods ethologists use when they observe animal behavior. You will work by yourself, with your group, and with your class to record and analyze your observations. You will then use those observations and what you know about the animals to explain why they are behaving the way they are.

Get Started

You will begin your investigations of animal behavior by observing familiar animals: middle-school students. Your teacher has asked a small group of your classmates to act as middle-school students usually act. They will show you about three minutes of behavior. They will try to act as they would in their natural environment. Your goal is to accurately observe and record the details of the students' behavior. Be sure to watch carefully and record your observations as specifically as you can.

Communicate

Three minutes of behavior might have seemed short when you first started watching it, but it is amazing how much there is to see in three minutes. Share your observations with your class. Describe what you saw as carefully as you can. Listen carefully so you can decide if all of you saw the same things or if you saw different things. Listen to the details of each observation so you can decide if the observation was accurate.

Discuss with your class the challenges of making observations. Describe the way you made your observations. Did you focus on one person or the whole group? Perhaps you listened more than you watched. Tell your class how you made your observations. Tell your class how confident you are of your observations.

Reflect

You probably saw many of the same things as your classmates. But you might have described them a little differently. Details are very important. Details help you tell the difference between accurate observations and inaccurate ones. Answer these questions to help you think about making observations and how you might improve on making observations.

1. After you listened to others' descriptions, how confident were you about your observations?

2. The next time you make observations, what will you do to make your observations as detailed as possible?

3. How important was the amount of detail people included in their observations? How did the level of detail help you know what happened in the middle-school scene?

4. Look back at your observations. If someone else read them, could they sketch a picture of what you saw? What could you do to help someone be able to sketch a picture?

What's the Point?

Scientists make observations to help them find out more about the world. When scientists make observations, they include as much detail as possible. When they include details, other scientists can read their descriptions and understand them. Then other scientists are able to decide if the observations are accurate. An ethologist is a scientist who studies how animals behave.

Eastern bongo antelope calves in their natural habitat. Their coats are reddish brown with bright white strips, which help provide camouflage from their enemies.

1.2 Explore

How Can You Improve Your Data Collection?

Plan Your Data Collection

Ethologists need to record what they see in a way that will allow them to remember it later. One way to record behavioral data is by using an **ethogram**, a table that describes behavior. Scientists also need to agree with each other about what they saw. They might want to compare what they saw one time to what they saw another time. They need good records to do that. They also need to record what they saw well enough so other scientists can use their data. In the previous section, you discussed some of the things that make it difficult to make good records of your observations. In this section, you will use a type of ethogram to keep records.

| **ethogram:** a table used to record observations of animal behavior. |

You will have another chance to observe and record the behaviors of middle-school students. This time you will use a type of ethogram to keep records. Work together with your group to plan how to keep good records of what you see. Tell each other what you think you saw the first time. Pay attention to where you agree and disagree with each other. Then, think about what you will have to watch and what you will have to record to be able to make detailed observations that you agree about. The questions below can be used to help you think about issues that might affect your observations. To make accurate and detailed observations, you might also think about whether others could draw a picture based on your description.
If they can, your description is very detailed. Consider the following questions:

- Will you watch an individual or all the students in the small group?
- How will you make sure you have observed all the members of the group?
- How will you make sure you have observed all the different behaviors?
- Will you each watch the whole scene or will it help to divide the observation task among all the members of your group?
- How will you record what you see?
- Will you take notes on everything you see?
- How can you record quickly enough so you don't miss anything? Can you write key terms and not full sentences?
- Will you draw a picture?
- Will you keep track of the amount of time each behavior lasts? If so, how will you do this?

After you have discussed your ideas with your group, develop a plan you will use to make your observations. Use the questions to help you decide on your plan. Make sure all members of your group agree on the plan and know how to follow through on it.

Observe

As you watch your classmates, pay careful attention to the details of the situation. Make sure you record all your observations accurately. Follow the plan you made in your group as closely as possible.

Analyze Your Data

The next step after collecting observational data is to make sense of it. You need to analyze it to understand what was going on. When scientists analyze observational data, they first check to see if their observations are the same as those of others. They discuss anything different and try to come to agreement. Next, they work to organize the data in a way that will let them see **categories** and **trends**. They identify the trends in the data—sequences of behaviors that they see over and over. You will do that with the observations you and your group have made. After you have completed the steps below, you will present your analysis to the class.

category: a set or class of things with similar characteristics, properties, or attributes.

trend: something that occurs over and over again.

1. Determine what everyone saw.

   Allow each group member to describe the behaviors they saw. Listen carefully to the observations of other group members. Decide if all of you saw the same things or if some of the observations were different. Think about why the observations might be different, and share your ideas with your group. Be sure to report any difficulties you might have had using the observation procedures your group decided on. It is important that you describe not only what you observed but also how you made each observation. This will help other students understand the data.

2. Organize your data into categories.

   The first step in analyzing your data is organizing it. You probably have many notes. You cannot understand your data with disorganized notes. Begin by copying each observation onto a sticky note. Put only one observation on each note. Your entire group will also copy their observations. If some are repeated, you can make just one sticky note for that observation.

Once each observation is on a separate sticky note, read through them and discuss how they might be grouped into categories by behavior. Combine observations of similar behaviors. If you disagree with the categories, ask others why they think a behavior fits into that category. Sometimes it is necessary to reorganize categories as more observations are read. Once you have each observation in a pile, use index cards to label the categories.

3. Identify trends.

As a group, prepare a description of the behaviors in each category. The description needs to be specific enough that other students in your class can recognize it. This will help you describe your groupings to the rest of your class.

Communicate

Each group will now share the categories they created and the reasoning behind their categories. Listen carefully to see if others made the same observations you made. Also, listen as each group describes their category labels. Think about whether they used the same labels as your group. Consider whether each group's labels would be understandable by other scientists.

Reflect

Each group probably did not create exactly the same categories, but each group had reasons for creating its particular categories. You will be learning more about observing animals and creating categories during this *Learning Set*. To prepare for that, answer the following questions and discuss the answers with your class. Listen carefully as other groups share their ideas. By working together, you may hear ideas that can help you improve your work. You will use the answers to improve your procedures, observations, and analyses.

Feeding is one category of behavior common to all animals, including humans.

1. What are the main categories of behaviors you found in your observations of the middle-school students?

2. How difficult was it to stick to the procedure you decided on in your planning?

3. What were the issues in your procedure that affected how you made your observations? Were your observations more complete or less complete because of your procedure?

4. What will you do to improve your procedure the next time you make this type of observation?

5. What will you do to improve your observations and analysis for the next time?

6. Why do you think different groups made different observations and identified different categories?

reliable data: data that is the same when collected many times or by different people.

Be a Scientist

Reliable Data

In this Unit, you will be collecting a lot of information. Most of the information will come from your observations of animals. This information is called data. Data can be collected in a lot of different ways. In some Units, you will collect data made up of numbers. You may set up an experiment and measure time, distance, or temperature. In this Unit, you will collect and analyze *observational* data.

Making good scientific decisions requires **reliable data**. Observational data is more reliable when the same procedure is used each time the data is collected. The same procedure also means that another person could use the procedure and collect similar data.

In this Unit, you will use the data from your observations to develop interpretations. Scientists use data to develop their interpretations. If the data are very reliable, then the interpretations are more reliable as well.

What's the Point?

Ethologists look for reasons why animals behave the way they do. They try to answer four different questions:

- How do different behaviors help an animal survive?

- How do the animal's environment and learning affect different behaviors?

- How do behaviors change as the animal grows?

- How do animals that are similar to each other act in similar or different ways?

To answer these questions, scientists need to make and analyze observations. Accurate, detailed observations are essential to understanding animal behavior. When you record your data, you can better analyze your data, share your work with others, and answer scientific questions. Ethologists share their observations and analyses with others who can help decide if the observations are accurate and detailed enough to help determine why animals behave as they do. They can do this only if they make careful observations, support their reasoning, and analyze their data to correctly identify categories of behaviors.

The behavior of these ducks could be in the category of parenting.

1.3 Explore

Observing and Interpreting Animal Behavior

You have made some observations of animal behavior by watching your classmates. While you were making your observations, you might have thought about how difficult it was to make accurate observations of animals. Observing animals' behavior can be difficult. Many things happen in a short time. Not only do you have to watch the animals themselves, but you have to be aware of what is happening around them.

You may have had difficulty sticking with your group's plan. You may have noticed that when you discussed your observations with your group, you had each made different observations. All these experiences show the importance of designing good plans for making observations if you want to understand animal behavior. Ethologists work hard to make accurate observations and use them to support their interpretations of animal behavior.

In this section, you will apply the observation methods you used in the last section to observe some animal behavior. It would not be possible to bring a lot of animals to your classroom, so you are going to observe pictures of animals to determine what the animals are doing.

You might observe that this dog is brown and has long ears. But, if you say that this dog looks guilty, you are interpreting the dog's behavior.

As you make your observations, be sure to look at all parts of the picture. Describe what the animal is doing in the picture. Pay attention to whether the animal is with others or alone. Take note of the animal's environment. Keep in mind the questions ethologists investigate, especially how the animal's environment affects its behavior.

Observe

Pictures of Human Behavior

Begin this investigation by looking at one of the pictures of people below. Use the same techniques you learned in the last section. Pay attention to the details in the picture. What is happening around the people? What knowledge can you gain from the picture? On your own, write a description of the picture. Record at least two behaviors you observe in the picture.

Picture #1

Picture #2

Picture #3

Picture #4

Conference

When you have completed your observations, discuss them with your group. Discuss what is happening in your picture and why you think the details you recorded are important. Could someone draw a picture based only on your description? If not, return to your picture and, with your group, write more descriptive observations. As you discuss your observations, work with your group to write descriptions on which you all agree. Make a list of these observations.

Communicate

Share your descriptions with the class. As you listen to one another's descriptions, notice where you agree and disagree. Think about how you might have described some of the pictures differently. Perhaps the class can identify some reasons why it is so hard to come to agreement about how to describe what you see.

interpretation: a description of the meaning of something.

Be a Scientist

Observation and Interpretation

When people say that animals are smiling, laughing, or talking to each other, they are giving animals human traits. Sometimes when people observe animal behavior, they describe the animal behavior as if the animals were people. Many people believe animals think and feel the way people do. It is natural for people to think animals have human feelings and emotions. They are interpreting the animals' behavior as though it is similar to human behavior. Thinking about animals as having human behaviors can become a problem when you study the behavior of animals scientifically. In a scientific study of animal behavior, it is important to observe what animals are actually doing. When ethologists record their observations, they work hard to separate observations and **interpretations**. They also make sure any interpretations they make are supported by many observations.

You make observations all the time. You notice that your best friend has a yellow shirt. You observe the teacher showing you how to do something. You look at a picture in a newspaper. You see the freckles on your arm. Descriptions of things you can see are observations. When you write accurate and detailed observations, someone else could almost draw a picture from your observations.

Interpretations are different from observations. When you make an interpretation, you take what you see, add what you know from previous experience, and decide what is happening. Interpretations are essential in science. Interpretations can help scientists better understand animal behavior. They can then use this understanding to predict what similar animals might do or what an animal might do in a particular situation. Scientists share their interpretations with other scientists and check to see if other scientists agree with the way the interpretations have been made.

Analyze Your Data

Use *Observing and Interpreting Animal Behavior* pages as you analyze your data. It will help you separate observations from interpretations. The page will also provide you with space to record details about the animals you are observing and their environments. These other details may help you interpret and answer the question about how the animal's environment affects its behavior.

Using this new page, separate your original list of observations into two lists, one of observations and one of interpretations. Put observations in the first column and interpretations in the last. Remember that an interpretation provides the meaning of or reason for an action. In the picture of the dog at the beginning of this section, the interpretation might be that the dog is guilty of something because of how it looks. Look through your group's observations. See if any of the observations you wrote were actually statements that show meaning or describe reasons for behavior. If so, write these in the *Interpretations* column.

Observing and Interpreting Animal Behavior

Name: _____ Date: _____

Animal I am observing: _____

Observations	What about the environment and animal allows that behavior?	Interpretations

You can pick out interpretations because they do not describe the picture, but instead describe an emotion, a look, or a reason why a person or animal is doing what they are doing.

Use the middle column to record information about why you think your interpretations are good ones. If you have made observations about the environment in the picture, list those in this column. If you know something about parents, teams, children, or anything else that helped you interpret, that goes into the middle column, too.

You will use the *Observing and Interpreting Animal Behavior* page many times in this Unit. It will help you track what you are observing, interpreting, and learning about animal behavior. The *Big Challenge* of this Unit is to develop a new enclosure for an animal using what you learn about the animal's behavior. The *Observing and Interpreting Animal Behavior* pages you create throughout this Unit will be helpful in keeping your observations and interpretations organized. You will then use this information to support your design for the final challenge.

Stop and Think

1. How difficult was it for you to support your interpretations of people in the picture you looked at? What knowledge did you use for those interpretations?

2. When you discussed one another's observations, what was confusing? What were your disagreements about the observations? About the interpretations? Why do you think these happened?

These baby birds are waiting for food.

Be a Scientist

Keeping Records

It is very important that scientists record their work carefully. To record means to write, sketch, or diagram what is being done. This allows scientists to accurately report their findings to others and helps them answer scientific questions.

You have probably kept records of your work in science class before. Keeping records is also important for you as a student scientist. Recording your work helps you:

- share your work with others,
- remember what you did and decided along the way,
- remember what you saw and the environment surrounding your observations, and
- answer scientific questions.

Throughout this Unit, you will be collecting a lot of observational data. Because of the amount of data you will collect, you will need to keep it well organized. Scientists frequently use tables to record and organize their data. Tables help scientists keep track of what they are seeing. The information in the table may include length of time a behavior happened, the surroundings of the animals, and perhaps even some interpretations.

Observe

Pictures of Other Animal Behavior

Now you will use what you have learned about making good observations to observe and interpret what is happening in pictures of other animals. Again, you will look carefully at one picture.

Recording Your Data

Describe the picture using good descriptive words. Record as many details as you can. Record your descriptions of each behavior on the *Observing and Interpreting Animal Behavior* page. This time, be aware of the difference between observation and interpretation. Use the correct column on the *Observing and Interpreting Animal Behavior* page to show which of your comments about the picture are observations and which are interpretations.

Picture #5

Picture #6

Picture #7

Picture #8

Conference

When you have completed your observations, discuss them with your group. Explain why you think details are important. Could someone sketch a picture based only on your description? If not, return to your picture and, with your group, write more descriptive observations. As you discuss your observations, work with your group to write descriptions you all agree upon.

Next, discuss your interpretations. Have group members read their interpretations. As you listen to others' interpretations, consider how their ideas might be different from yours. Together with your group, determine why the interpretations are similar or different.

Communicate

You will now share your observations and interpretations with the class. As you look at your classmates' pictures, think about whether you would have made the same interpretations. If you have a different interpretation or you do not understand a group's interpretation, you should ask questions to help you understand. Then discuss these questions.

1. You observed and interpreted two pictures, one of a human and one of another animal. Which of the two pictures did you find easier to observe? In which picture was it easier to interpret the behavior? Why do you think this is?

2. How did using the *Observing and Interpreting Animal Behavior* page make it easier to record your observations?

3. What is the difference between observation and interpretation? You might use a new example of both to show the difference.

4. Why is it important for scientists to think about the difference between observation and interpretation?

5. Why do you think it is sometimes easy to confuse observation and interpretation?

What's the Point?

Ethologists observe animal behavior. They watch and record what animals do. During their observations, they collect a lot of data and often use a table to keep track of the data. By keeping good records, ethologists can re-create what they have seen and use that information to determine why animals behave as they do.

Interpretation is different from observation. Interpretation includes the meaning or significance of an action. It is difficult, when making observations, to eliminate all interpretation. However, scientists try to separate their interpretations from their observations. Sharing ideas with others can also make separating observations and interpretations easier.

1.4 Explain

Support Your Interpretation

You have made accurate observations and learned how to separate observations from interpretations. When you were making observations of the animals in the pictures, you learned that it is difficult not to interpret behaviors as you are describing them. During your discussions with your team and the class, you were asked why you made your interpretations. They were asking you to support your interpretations with evidence.

explanation: a statement that connects a claim to evidence and science knowledge.

Now you will more formally connect your observations and interpretations to each other. You will write **explanations**. An explanation is a statement that connects a claim (a conclusion you have come to) with your evidence (the data you have collected) and science knowledge. The explanations you write will connect your observations and what you know about animals to your interpretations.

> ### Be a Scientist
>
> ## What Do Explanations Look Like?
>
> Making claims and creating explanations are important parts of what scientists do. An explanation is made up of three parts:
>
> **Claim**—a statement of what you understand or a conclusion that you reach from an investigation or set of investigations.
>
> **Evidence**—data collected during investigations and observations.
>
> **Science knowledge**—knowledge about how things work. You may have learned this through reading, talking to an expert, discussion, or other experiences.
>
> An explanation is a statement that connects the claim to the evidence and science knowledge in a logical way. A good explanation can convince someone the claim is valid.
>
> For example, suppose you have a pet hamster. You notice that when you remove the lid to the cage to feed the hamster, it runs toward the food cup. You learned in science class that some animals learn different behaviors. Dogs who are fed after a bell rings eventually respond to just

the bell. They begin to drool when they hear the bell, whether or not food is offered to them. You wonder if the hamster might have learned that the noise of the lid signals it is about to be fed. The next time you feed your hamster, you try to remove the lid very quietly and notice your hamster does not move toward the cup. You conclude that your hamster has learned that, when the lid makes noise, the food cup will be filled, and it is time to eat. You can now form an explanation.

Your claim: My hamster moves toward the food cup because of the noise of the lid.

Your evidence: I performed an experiment. My experiment showed that when there is no noise, the hamster does not move. The hamster moves toward the food cup when I lift the lid normally (with noise). When I lift the lid quietly, the hamster does not move toward the food cup.

Your science knowledge: Some animals, like dogs, can learn different behaviors.

Your explanation: My hamster moves toward the food cup because of the noise of the lid. When it hears the noise of the lid, my hamster moves toward the food cup. This is because my hamster has learned that when it hears me lifting the lid, I am going to feed it.

An explanation is what makes a claim different from an opinion. When you create an explanation, you use evidence and science knowledge to back up your claim. Then people know your claim is not simply something you think. It is something you have spent time investigating. You have found out things that show why your claim is likely to be correct.

Explain

Now that you know more about what an explanation is, you are going to write an explanation of the animal behavior you just observed. Use a *Create Your Explanation* page to help you make sure your explanation takes into account your claim, evidence, and science knowledge. Your interpretation of the animal's behavior is your claim. Your observations of behavior and what you recorded about your animal's structure and environment are your evidence. You may have some science knowledge from your own experiences or from readings. Record all of these in the appropriate boxes.

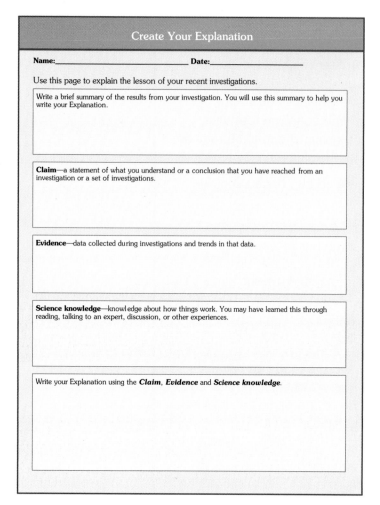

Create Your Explanation

Name:_____ Date:_____

Use this page to explain the lesson of your recent investigations.

Write a brief summary of the results from your investigation. You will use this summary to help you write your Explanation.

Claim—a statement of what you understand or a conclusion that you have reached from an investigation or a set of investigations.

Evidence—data collected during investigations and trends in that data.

Science knowledge—knowledge about how things work. You may have learned this through reading, talking to an expert, discussion, or other experiences.

Write your Explanation using the *Claim*, *Evidence* and *Science knowledge*.

Then write a statement using your evidence and science knowledge to support your claim. This is your explanation. A good explanation can convince someone else that your interpretation is good. If your statement doesn't seem convincing, revise your claim so your evidence and the science you know will support it. You can use the hamster example to know what to put in each part of your explanation.

Because your understanding of the picture you observed may not be complete, you may not be able to fully explain the animal's behavior. But use what you have read and what you know to develop your best explanation. Scientists finding out about new things do the same thing. When they only partly understand something, it is impossible for them to form a "perfect" explanation. They do the best they can based on what they understand. As they learn more, they make more accurate or clearer explanations.

This is what you will do now and what you will be doing throughout PBIS. You will explain your results the best you can based on what you know. Then, after you learn more, you will make your explanations more accurate.

Communicate

Share Your Explanation

Now you will share your observations, interpretations, and explanations with the class. Your classmates will listen carefully. They will want to know what your claim is. They will be checking to see if your claim is **valid**. You will have to make sure you support your claim with accurate and detailed observations and appropriate science knowledge. If you have done that, then your claim can be considered valid.

valid: well-grounded or justifiable.

Reflect

Explanations are critical for scientific understanding. Scientists in all science fields are trying to explain the way the world works. They develop explanations to help them inform other scientists of what they have learned from their observations.

1. After scientists make their observations and interpretations, they write explanations of what they have seen. Creating explanations can be difficult. What difficulties did you have developing an explanation of what you saw?

2. What questions do you still have about how explanations can be made?

3. One source of science knowledge might be your own experiments. List three more sources you might use for science knowledge.

What's the Point?

Science is about understanding the world around you. Scientists gain understanding by making observations and explaining what they see. Scientists make claims about what they observe. They support their claims with evidence they gather through their observations. They also look at the data, claims, and explanations others scientists have published. They combine all of that to create explanations. Other scientists carefully examine these explanations. They discuss them with each other and try to determine if the claim is valid. Valid claims require good observations and science knowledge as support.

Throughout PBIS, you will create explanations. Every explanation will include a claim, evidence, and science knowledge. Just like scientists, you will edit and improve your explanations.

To help others better understand what they learn, scientists must communicate their results effectively. When scientists share what they have learned, they allow others to question it and improve on the claims and explanations. By working together, scientists can develop a clearer understanding of the world.

1.5 Read

What Do Animals Need to Survive?

behavior:
an animal's
response to its
environment.

adaptation: a
special trait that
allows an animal
to survive in its
environment.

species: all of
living things of
one distinct kind
with common
characteristics,
such as dogs.

**instinctive
behavior:** a
behavior an
animal is born
with.

**learned
behavior:** a
behavior that
comes from
teaching or
experience.

mammal: a
warm-blooded
animal with hair
and in which
the female has
special glands to
feed milk to its
offspring

In this *Learning Set*, you have been exploring how scientists study animal behavior. One of the questions you are answering in this Unit is *Why do animals behave the way they do?* To answer that question, you need to understand what a **behavior** is from a scientific point of view. Scientists use the word behavior to describe an animal's response to its environment. One of the most important animal behaviors is survival.

Animals share certain needs for survival. These needs affect how they behave. They must obtain enough food for energy and to feed their young. They need protection from bad weather and enemies, and they need a safe place to raise a family. By using behaviors and other **adaptations**, or traits, to meet these basic needs, not only can an animal survive, but its family may survive, its group may survive, and its entire **species** may survive.

Lions live, travel, and hunt in groups called prides. This allows them to hunt cooperatively to catch bigger prey more often and protect each other.

Kinds of Behavior

A behavior can be **instinctive** or **learned**. Animals are born with instinctive behaviors. Learned behaviors come from teaching or experience. Most behaviors are a mixture of instinctive and learned. For example, **mammal** mothers have the instinct to care for their young, but some then learn from experience how to better care for their young. Whether a behavior is instinctive or learned, the most important thing an animal must do in its environment is survive.

AIA 32

Searching for Food

Animals need energy and materials to stay alive, grow, develop, and reproduce. They obtain the energy and the nutrients they need from the food they eat. The process by which animals break down their food to obtain energy and then use the energy is called **metabolism**. Finding enough of the right kind of food depends on many things: how and where animals live, how they find and gather or catch food, and how they digest that food.

An animal must live in a place that has the type of food it needs and it must have ways to gather or catch the amount of food it requires. Some animals have claws and big teeth; others are able to run fast. Animals might live in cooperative groups and help one another find food. Other animals live alone.

An animal must be able to digest the food it eats. Even if an animal finds a lot of food that has a lot of energy, it cannot use the energy if it cannot digest the food. Energy in food is measured in units called **calories**. A calorie is the amount of energy needed to raise the temperature of one gram of water by 1°C. Most foods contain thousands of calories of energy. For this reason, scientists use the **Calorie**, with a capital *C*, to measure the energy in foods. One Calorie is the same as 1 kilocalorie or 1000 calories.

Some animals eat only meat, and some eat only vegetable matter. Some eat both. All have different ways of getting their food and digesting it. You will learn more about feeding behaviors and adaptations in *Learning Set 2*.

Protection

In addition to finding food, animals must remain safe—from their enemies and from their external environment. Remaining safe from enemies is a primary concern for many animals. Organisms have developed different behaviors to keep them safe. Birds may form **roosts** that can number in the thousands. With many birds together in one place, many eyes are on the lookout, and each bird is safer than if it were sitting on a branch all alone. Some mammals that live in herds form a protective ring around their young if they are threatened by predators. Others like turtles retreat into their hard shells when they sense danger.

metabolism: the combination of chemical reactions through which an organism builds up or breaks down materials converting energy to carry out its life processes.

calorie: the amount of energy needed to raise the temperature of one gram of water by 1°C.

Calorie: the amount of energy in foods. One Calorie is the same as 1 kilocalorie or 1000 calories.

roosts: communal resting places, mostly for a single species of birds.

Some birds form roosts in trees at night for protection from enemies.

ANIMALS IN ACTION

Many animals living in cold climates, such as these polar bears, have fur, fat, and short limbs to conserve body heat.

Some animals, such as this golden retriever, pant to cool off their bodies.

But even if they are safe from enemies, animals might not be safe from their environment. Animals need a way to keep their internal temperature and fluids fairly constant. This process is called **homeostasis**. Some animals that live in hot areas have a special circulatory system that moves cooler blood across the brain, so they will not die from heat stroke while running. Or they may sweat or pant to cool off their bodies. Animals may shiver if they get too cold. The muscle action used to shiver produces heat that warms their bodies.

Reproduction

If animals have enough food and are safe, they can use their energy for reproducing and raising their young. Animals raise offspring in one of two ways. Either they feed and protect the young until they are old enough to go out on their own, or the mother leaves the young to manage for themselves.

homeostasis: the maintenance of stable internal conditions in an organism.

mammary glands: milk-producing glands found in female mammals that are used to feed the young.

One characteristic of mammals is that the mother raises the young on mother's milk. She produces the milk in her **mammary glands** and continues to feed her young until they are old enough to eat on their own. A mother must obtain enough energy from her food to meet her own needs and to produce milk for her young.

In many birds, both the mother and the father help feed and protect the young, which are called **hatchlings**. In some birds, the parents gather food for the young, process it in their own stomachs with **enzymes**, and then **regurgitate** it to feed to the young. These parents continue to feed their offspring until the young birds can gather food on their own.

Other animals, like fish and turtles, lay many eggs in the water or on land, and then leave. When the eggs hatch, the young must take care of themselves.

Why do animals use different strategies for raising their young? In animals that rely on learned behaviors for survival, it may take a long time for the young to learn how to find food, how to find shelter, and how to protect themselves. In this case, the mother or both parents stay with and protect the young while they teach them what they need to know. In animals that rely mostly on instinctive behaviors, the young are born with the instincts they need to survive and do not need their parents.

hatchling: a very young baby bird.

enzymes: organic substances that cause chemical changes in other substances.

regurgitate: to bring partially digested food up from the stomach into the mouth.

A female turtle lays eggs, but she does not care for her eggs or the hatchlings. The hatchlings are born with the instintive behavior they need to survive. A female spider dies in the autumn, before the spiderlings (baby spiders) hatch the following spring. However, the spiderlings have the instinct to build a perfect web on their first attempt.

Many birds, such as these penguins, take care of their young and feed them regurgitated food.

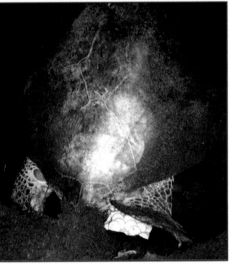

Mammals produce milk in mammary glands to feed their young (left).

Some animals, such as this sea turtle, lay many eggs but leave the young to manage for themselves (right).

ANIMALS IN ACTION

Stop and Think

1. Some large animals, such as lions, live and hunt in groups. What are some of the advantages of hunting with others?

2. Living in a group can help animals with protection. Describe two ways that living in a group can protect individuals.

3. Some animals have only a few young at a time. Others have many offspring, even thousands. How do the numbers of offspring connect with how the animals care for their young?

Revise Your Explanation

You have just read about how an animal's behavior and adaptations help it survive in its environment and carry out its basic needs. With your group, look at the explanation of animal behavior you created on your *Create Your Explanation* pages after interpreting the pictures of animals. Now that you know more about the science of animal behavior, you can probably revise your explanation based on some of this science knowledge.

Review and rewrite your explanation based on your new science knowledge. First, check to make sure your claim is accurate. You may have just read information that shows that your claim was inaccurate. If your claim does not match the science you have read, revise it. Next, support your claim with the science knowledge you just learned.

Then, rewrite your explanation to make it more complete. Remember that an explanation is a statement that connects a claim to evidence and science knowledge in a logical way. Try to write your explanation so that it tells why your claim is true. Be sure that your explanation matches the science you just read.

Communicate

Share Your Explanation

Share your new explanation with the class. When you share your explanation, tell the class what makes this revised explanation more accurate than your earlier one. As each group shares their explanation, pay special attention to how the other groups have supported their claim with science knowledge. Ask questions or make suggestions if you think a group's claim is not as accurate as it could be, or if the group has not supported their claim well enough with observations and science knowledge.

Reflect

Explanation is an important scientific practice. Scientists use what they already know along with new evidence collected from investigations to explain how the world works. You will be doing a lot of scientific explanation in PBIS. It will get easier as the year goes on. For now, think about the differences between the first animal behavior explanations your class wrote and the new explanations. Identifying what makes the earlier explanations different from the newer ones will help you get better at explaining. Answer the following questions and be prepared to discuss them in class.

1. What are you able to explain now about the animal behaviors in the pictures that you were not able to explain well earlier?

2. What makes your revised explanations better than the earlier explanations?

3. In order to make a complete explanation, science claims need to be supported by evidence. What are some sources of evidence you might use to support your claims?

Be a Scientist

Good Explanations Tell Us How and Why

A good explanation uses what scientists know about how things work. The best scientific explanations use agreed-upon science knowledge in a logical way to support a claim. These kinds of explanations can usually convince others that a claim is valid. And with more science knowledge, you can write better explanations.

What's the Point?

You have learned that animals have many different behaviors and adaptations to help them survive in their environments. These behaviors can be instinctive or learned. Animals gather or catch food to obtain energy. They use other behaviors and adaptations to remain safe from enemies and to protect themselves against extreme weather. When energy gathering is efficient and animals are safe, they can reproduce. Animals care for their young in many ways, from total care and protection to laying eggs in water or on the ground and leaving the young to manage for themselves. Survival and success in animals depends on a wide variety of behaviors and adaptations.

Learning Set 1

Back to the Big Question

How do scientists answer big questions and solve big problems?

The *Big Question* for this Unit is *How do scientists answer big questions and solve big problems?* You are answering the *Big Question* in the context of answering the science question: *Why do animals behave the way they do?* You will be applying your answers to those questions to a *Big Challenge*, creating an animal enclosure that allows an animal to behave as it would in its natural environment. To design an enclosure that meets all the criteria, it is important to know about how the animal behaves in the wild and what affects that behavior. You have learned that to better understand why animals behave as they do, scientists make observations. You have developed and used procedures for observing animal behavior. You will use those procedures as you learn more about animal behavior and address the *Big Challenge*.

iteration: a repetition that attempts to improve on a process or product.

Earlier in this *Learning Set*, you designed an observational plan to watch your classmates. Now that you've learned so much more about observing animal behavior, you will develop a new plan. It will probably make your observations a lot easier and allow you to collect more reliable data. Each time you develop a plan, you can use the successes and challenges from your previous experience to make a better plan. Scientists use their previous experiences to improve their procedures too. When they improve their procedures, the data they collect are usually more reliable. Scientists often perform almost the same investigation over and over again, each time using what they have learned earlier to make their next investigation and data more reliable. When someone redesigns a procedure or product based on what they have learned, it is called **iteration**. The word refers to the process of revising a plan or product. We also refer to each revision as an iteration.

Be a Scientist

Iteration

Scientists develop procedures for their investigations. They try to make sure procedures have the right steps that are followed in the right order. But sometimes procedures do not work out as they had planned. When that happens, scientists change their procedures based on where they might have had difficulty and run the investigation again to collect better data. They are continually learning from their investigations.

In your group, you wrote a procedure the first time you observed your classmates. There were things that worked well with your procedure and things that needed improvement. Iteration gives you a chance to revise the procedure and not make the same mistakes again.

You will now use all the things you have learned about how ethologists design their plans, make their observations and collect data, organize and analyze data, and describe their results, to develop a new observation plan. Then you will observe the students in your class one more time.

Later you will use the skills and processes you are learning to study animal behavior. You will use the observations you make of different animals to understand why animals behave the way they do. Once you understand that, you will be able to apply your understanding to designing an animal enclosure.

Plan

You will continue your investigations of animal behavior by observing your classmates again as they are engaged in an activity. Your goal is to accurately observe and record the details of the students' behavior and to interpret their behavior, explaining why they are doing what they are doing.

With your group, revise your plan to make more accurate observations. Make sure all members of your group agree on the plan and know how to follow through on it.

- Will you watch an individual or all the students in the small group?

- How will you make sure you have observed all the members of the group?

- How will you make sure you have observed all the different behaviors?

- Will you each watch the whole scene or will it help to divide the observation task among all the members of your group?

- How will you record what you see?

- Will you take notes on everything you see?

- How can you record quickly enough so you don't miss anything? Can you write key terms and not full sentences?

- Will you draw a picture?

- Will you keep track of the amount of time each behavior lasts? If so, how will you do this?

Observe

As you watch your classmates, pay careful attention to the details of the situation. Follow the plan you made in your group as closely as possible. Record your observations as accurately as possible. Try to observe everything about the scene, your classmates, and the people around them. Watch what they are saying and what they are doing.

Analyze Your Data

After the observation time is over, discuss your observations with your group. Allow each group member time to describe the behaviors they saw. Decide if all of you saw the same things or if some of the observations were different. Use the same procedure you used before to analyze your group's data.

The first step is to copy each observation on to a separate sticky note. Put only one observation on each note. Your entire group will also copy their observations. If some are repeated, you can make just one sticky note for that observation.

Once each observation is on a separate sticky note, read through them and discuss how they might be grouped into categories by behavior. Combine observations of similar behaviors. If you disagree with the categories, ask others to be clear about why they think a behavior fits into that category. Sometimes it is necessary to reorganize categories as more observations are examined. Once you have each observation in a pile, use index cards to label the categories.

As a group, prepare a description of the behaviors in each category. The description needs to be specific enough so that other students in your class can recognize it. This will help you describe your groupings to the rest of your class.

Then, using an *Observing and Interpreting Animal Behavior* page, record your observations of the students' behavior, your interpretations of their behavior, and anything important about the environment and the students that helped you make each interpretation.

Communicate

Each group will now share the categories they created, the reasoning behind their categories, and the behaviors and interpretations they recorded. Listen carefully to see if others made the same observations you made. Also, listen as each group describes their category labels. Think about whether they used the same labels as your group. Consider whether their labels would be understandable by other biologists.

Think about how the categories are the same as or different from the categories you created earlier in the unit. The students you observed were behaving differently, so the categories may be very different. It is interesting that some of the categories might also be the same, even though the overall behavior was very different.

Reflect

Iteration is an important part of scientific work. Repeating the same procedures again helps to check the validity of observations. Iteration also gives you a chance to revise your procedure when you learn it is difficult to do or when parts are missing.

1. What did you learn from revising and retrying your procedure?

2. In what ways were the observations and interpretations easier now than they were at the beginning of this *Learning Set?*

3. In what ways are your observations and interpretations more reliable? What did you do differently that made them more reliable?

4. What do you think you might do to make the observations and interpretations easier and more reliable next time?

5. How do you think iteration will help you when you design your animal enclosure?

What's the Point?

When ethologists study animal behavior, they have ideas about what they should do and how they should do it. They work hard to gather reliable data. One way to gather reliable data is to have a procedure to follow. One way to improve the validity of data is to revise a procedure when it does not work well and then collect new data. In the first section, you used a procedure, and in this section, you were able to update it. This revision probably helped you gather more reliable observations.

Learning Set 2

What Affects How Animals Feed?

The *Big Question* for this Unit is *How do scientists answer big questions and solve big problems?* In *Learning Set 1*, you observed some animal behavior. You also learned some skills ethologists use when they observe animal behavior.

What types of behaviors are ethologists interested in observing? They are most interested in looking at behaviors that are common to all living things. These behaviors include feeding, communicating, moving, playing, and taking care of their young. Each of these behaviors is very important to the life of the animal. It could take years to investigate each of these behaviors. Ethologists spend their lives observing animals. They use their collected data to make interpretations of what they see.

In this *Learning Set*, you are going to look at one animal behavior, feeding. You will observe several different animals feeding. You will answer the smaller question: *What affects how animals feed?* Through investigating, reading, and observing videos, you will begin to interpret feeding behaviors. You will need to consider the influence of the animal's body shape, size, and function on how it feeds. You will also observe each animal's environment to determine the effect of the environment on its feeding behavior.

Scientists can learn a lot by observing how animals feed.

2.1 Understand the Question

Thinking about What Affects How Animals Feed

herbivore: an organism that eats only plants.

carnivore: an animal that eats only meat.

omnivore: an organism that will feed on many different kinds of food, including both plants and animals.

forage: to search for something, especially food and supplies.

forager: an organism that searches for food.

predator: an organism that captures and eats part or all of another organism.

Feeding is one behavior common to all animals. From the smallest to the largest, from the least complex to the most complex, each animal needs food to provide it with energy.

Different animals eat different things. Some animals are **herbivores**. They only eat plant material. Some examples of herbivores are elephants, giraffes, squirrels, and bees. Other animals are **carnivores**. They only eat other animals. Lions, cheetahs, snakes, and spiders are carnivores. Still other animals are **omnivores**. Omnivores eat both plants and animals. Examples of omnivores are chimpanzees, bears, chickens, and flies.

Different animals also have different ways of feeding. Some animals **forage** or move from place to place looking for food under rocks or in the bark of trees. Many insects, including bees, are **foragers**. They move from plant to plant to find food. Many herbivorous animals (animals that are herbivores), such as giraffes and elephants, are also foragers.

A **predator** is an animal that eats other organisms. Some carnivores, like lions and tigers, are predators that hunt. Unlike foragers, these carnivores must move very quickly because their prey also moves quickly. Other predators hunt by waiting until the prey comes to them. For example, alligators are also carnivores, but they hide until their prey comes near, and then they pounce.

Scientists are interested in what animals eat and how they find food. They often observe animals while they are feeding to better understand how animals fit into their environment.

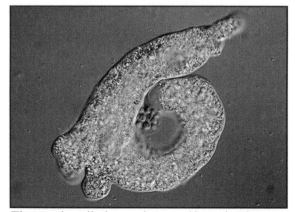
The single-celled amoeba engulfs its food.

The giraffe strips the leaves from trees with its powerful lips and long tongue.

You will need to learn more about how animals feed for another reason too—so you can address the challenge of designing an enclosure for one of the animals you are studying. Remember that the habitat will have to allow your animal to feed as it does in the wild and will have to allow scientists and visitors at the zoo to observe your animal as it feeds.

Get Started

You will watch a short video of chimpanzees feeding. Before watching the video, discuss with other members of your group what you know about chimpanzees. Try to imagine what the life of a chimpanzee might be like. Use the following questions to guide your discussion.

- Where do chimpanzees live?

- What do chimpanzees eat?

- How do chimpanzees get their food?

Listen to others in your group as they share what they know about chimpanzees.

Watch the video. Pay attention to the feeding behavior of the chimpanzees. Try to figure out what the chimpanzees are doing. Observe what they are eating. Notice if all the chimpanzees are eating the same way.

Conference

Discuss with your group what you saw in the video. What surprised you? What did you not understand? What do you disagree about? Identify what you think you know about how animals feed. Remember your challenge—to design an enclosure for an animal. What more do you need to learn about chimpanzees to be able to design an enclosure for chimpanzees? What might you need to learn about other animals to be able to design enclosures for them? Watching the chimpanzees feeding might have reminded you of some of the things you know about how other animals feed and some of the things you have wondered about in the past.

Working by yourself, develop two questions that might help you better understand what affects how animals feed. During earlier discussions with your group, you might have found that you disagreed about some things. As you identify what else you need to learn about how animals feed, keep in mind what you disagreed about, what surprised you, and what you did not understand. When you write your questions, keep in mind that your questions should

- be interesting to you,
- require several resources to answer,
- relate to the *Big Question* and designing a new enclosure that will encourage the feeding and communication of an animal, and
- require collecting and using data.

Make sure your questions are not simply yes/no questions or ones you can answer with a single word or sentence.

When you have completed your two questions, meet with your group. Share all the questions with each other. Carefully consider each question and decide if it meets the criteria for a good question. With your group, refine the questions that do not meet the criteria. Choose the two most interesting questions to share with the class. Give your teacher the rest of the questions so they might be used later.

Update the *Project Board*

Recall that the *Project Board* helps you organize your ideas as you answer the *Big Question* and address the *Big Challenge*. You will now share with the class what you think you know about how animals feed and your group's two questions. Be prepared to justify why yours are good questions. Your teacher will add your questions to the *Project Board*. Throughout this *Learning Set*, you will work to answer some of these questions.

What's the Point?

All animals must eat to obtain the energy and nutrients they need to live. Therefore, feeding is a behavior common to all animals. What animals eat and what affects how they feed can be determined through careful observation. Some animals eat meat (carnivores), some eat plants (herbivores), and other animals eat both (omnivores). When animals feed, they do it in a variety of ways. Some animals that are predators move quickly to chase and capture their prey, while other predators wait until their prey comes near them, and then they pounce. Some animals, like chimpanzees, are foragers. When foragers eat, they move from place to place looking for food.

2.2 Explore

What Affects How Chimpanzees Feed?

You have seen chimpanzees feeding. Ethologists look for answers to why animals behave as they do. Now, it is time for you to begin looking for these kinds of answers. In this section, you will work with your group, watch the video again, and begin to look at the factors that might affect how chimpanzees feed.

Observe

You will now watch the video again. This time, you will make observations that will help you answer the question for this section. Start by making an observation plan with your group. How will you go about observing the chimpanzee group in this video? Remember that when you observed middle-school students' behavior earlier in this Unit, you developed a plan. You might use a similar plan this time. Create your plan, keeping in mind that you need to observe as much of the scene as possible, and that your observations are going to determine how well you understand the feeding of chimpanzees. Make sure that part of your plan includes using a page to record your notes.

Once you have your plan written, and all the members of your group understand the plan, your teacher will start the video again. Remember to take notes about the chimpanzees' behavior and habitat. Pay attention to the chimpanzees' bodies. Think about how their body structure helps them feed the way they do. Remember, too, to pay attention to how the chimpanzees' habitat affects how they eat. Remember to follow the plan your group decided on. Record your observations so you will be able to share them with your group and your class.

Analyze Your Data

You may have had difficulty making your observations. Each member of your group may have seen something different. By sharing your observations with others, you might have discovered things that you did not notice before. You may also have had a difficult time working with the plan you created.

Meet with your group to share your observations. Each member of the group should have a chance to read all of their observations to the rest of the group. Depending on your observation plan, you may have been watching different things than other group members were watching. Listen carefully for observations you might not have seen. Create a group list of observations that everyone agrees on. Make a second list of observations that were not agreed on. Make a third list that includes anything that was difficult about following your observation plan.

Using the same procedure you used in *Learning Set 1*, create a sticky note for each observation on which you agreed. Organize your observations into groups based on the type of behavior shown. Most observations will be about feeding. The important thing here is to provide as much detail as possible about the chimpanzees' feeding activity. For example, one group of observations might be about how the chimpanzee gets ready to eat. Another might be about how they use tools. If you think some behaviors might fall into two categories, you can list those behaviors more than once. Just make a copy of the sticky note so you have one for each category that a behavior fits into.

Communicate Your Results

Investigation Expo

You will share your observations and data analysis in an *Investigation Expo*. Each group will make a poster showing what they observed and the way they grouped behaviors. Then each group will present their poster to the class. Remember that each group created their own plan for making their observations and analyzing their data. Each group's observations may have been affected by their plan. Different groups' analyses may be different from each other.

For your presentation, create a poster that describes each of the following:

- questions you were trying to answer in your observations

- your observation procedure and how it helped you make observations

- what you observed

- the categories of behavior you identified, what behaviors fit into each category, and how confident you are about this analysis

One way you might present your questions, observations, and data analysis is in the form of a diagram. Place each group of observations in columns on your diagram. You can use the example on the next page to help you.

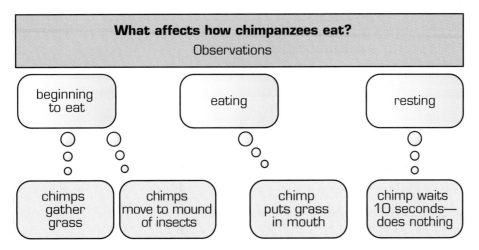

What affects how chimpanzees eat?
Observations

| beginning to eat | eating | resting |

chimps gather grass

chimps move to mound of insects

chimp puts grass in mouth

chimp waits 10 seconds— does nothing

As you present, be sure to be very detailed in presenting everything on your poster—your questions, your observation plan, your observations, your way of grouping those observations, how well you think your plan worked and why, and how much you trust your analysis.

As each group is presenting their poster, listen for the answers to the questions below. If a group does not answer all these questions, ask them questions to help you better understand what they found out and what their observation plan was. Remember to be respectful when you ask your questions.

- What was the group trying to find out?
- What procedure did they use to collect their data?
- How well were they able to make clear and detailed observations?
- How well did they group their observations? Does the way they grouped them make sense to you?
- What conclusions do their results suggest?
- Do you trust their observations? Why or why not?

Investigation Expo: a presentation of the procedure, results, and interpretations of results of an investigation.

Be a Scientist

Introducing an Investigation Expo

An *Investigation Expo* is like other presentations you have done. However, it is specially designed to help you present results of an investigation. Presentations during *Investigation Expos* include your procedure, your results, and your interpretations of your results. *Investigation Expos* are similar to presentations scientists make to each other. Scientists present results of investigations to other scientists. This lets other scientists build on what they learned. You will do the same thing.

There are several things scientists usually want to know about investigations. These include the following:

• questions you were trying to answer in your investigation

• your predictions

• your procedure and what makes it fair

• your results and how confident you are about them

• your interpretation of the results and how confident you are of it

To prepare for an *Investigation Expo*, you will usually make a poster that includes all of the items listed above. You must present those things on your poster in a way that will make it easy for someone to follow. Others should be able to identify what you have done and what you found out. If you don't think your procedure was as fair or as accurate as you had planned, your poster should also have a report on how you would change your procedure if you had a chance to run the investigation again.

Sometimes, scientists make posters when they present their investigations and results. They set up their posters in a large room where other scientists have also set up their posters. Then other scientists walk around the room. They look at the posters and talk to the scientists who did the investigations. Another way scientists share results is by making presentations. For presentations, they stand in front of a room of scientists. They talk about their investigations and results. They often include visuals (usually slides) showing all the important parts of their procedures and results. They talk while they show the visuals. Then other scientists ask them questions.

Your *Investigation Expos* will combine these practices. Sometimes, each group will formally present their results to the class. Sometimes, each group will put their poster on the wall for everyone to walk around and read. In this *Investigation Expo*, because every group's observations were a little different, each group will present to the class.

There are two parts to an *Investigation Expo*: presentations and discussions. As you look at posters and listen to other groups present their work, look for answers to the following questions. Make sure you can answer this set of questions about each investigation:

- What was the group trying to find out?
- What procedure did they use?
- Was it a procedure that could help them answer their question?
- Was it a procedure that allowed them to collect accurate data?
- How consistent is their data?
- What did they learn?
- What conclusions do their results suggest?
- Do you agree with their results? Why or why not?

When looking at posters and listening to presentations, you should ask questions if you cannot identify a clear answer to any of the questions above. Ask questions that you need answered to understand results and to satisfy yourself that the results and conclusions others have drawn are trustworthy. Be sure that you trust the results that other groups report.

Reflect

Each group noticed some different things about chimpanzee behavior. By sharing your observations and interpretations with others, you probably discovered things you did not notice when you observed the chimpanzees. Answer the following questions. Be prepared to discuss your answers with your group and the class.

1. In what ways were the chimpanzees different from what you thought they would be?

2. The chimpanzees used different parts of their environment to help them feed. Describe how you saw the chimpanzees using the environment while they were feeding. What would be different if this part of the environment changed? How do you think the chimpanzees would react?

3. What parts of the chimpanzees' bodies were useful as they were feeding? How were these parts like human parts? How were they different? How would feeding be different if the chimpanzees' bodies were built differently?

4. The chimpanzee video was shot in the wild. Why do you think it is important for scientists to study animals in their natural environment?

What's the Point?

Scientists often observe animals in their natural habitat. When they observe animals, they take detailed notes. They pay attention to everything happening around the animal. But it is still difficult to make good observations in the wild, because things are constantly moving and the action is fast.

Chimpanzees are omnivores. They eat both plants and animals.

2.3 Read

How Do Chimpanzees Feed and Why?

One amazing scientist of our time is Jane Goodall. She is famous for studying chimpanzees in the Gombe, a wild, natural place in Africa. As a young woman, Jane Goodall became the student of another famous scientist, Louis Leakey. He suggested that there was important scientific work to be done observing the behavior of chimpanzees in Africa. Jane Goodall thought this sounded like quite an adventure. Over 35 years have passed, and Jane Goodall is still working with chimpanzees. She is now sharing her knowledge with her own students and the rest of the world.

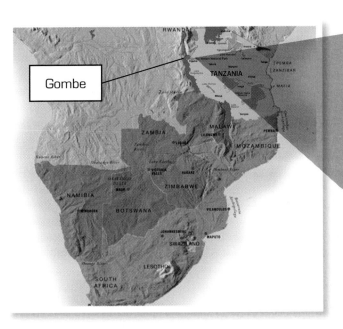

Gombe

Be a Scientist

Field Observations

Imagine you are a young scientist. Your name is Jane Goodall. You are working on your first scientific assignment on your own. That assignment is to observe the behavior of chimpanzees in their natural habitat in an area called Gombe in Africa.

It took months of patient observations from a distance before Jane Goodall could get close to the chimpanzees.

termite: an insect similar to an ant that lives in large colonies. Termites feed on wood and can damage wooden structures.

You have prepared for months for this new position. You read about Africa so you would know what to expect there. You read as much as you could about chimpanzees to help you know what others before you had learned about their behavior. You have been interested in animals since you were a child. You are anxious to learn about chimpanzees in the wild. However, soon after you start your observations of chimpanzees, you discover that your job is not as easy as you may have thought. The chimpanzees do not let you get close enough to observe them.

You think that perhaps the chimpanzees are afraid because they do not know you. You decide to be patient. You start watching them from afar. You hope they will not be afraid of you. You follow them for days as they walk in the forest. You watch the chimpanzees as they play, eat, and socialize. After a long while, they allow you to get closer, and they actually approach you.

One rainy October day, you see something extraordinary. You are walking through tall, wet grass. Ahead of you is a chimpanzee you recognize. You have named him David Greybeard. He is hunched over a **termite** nest poking the entrance of the nest with a blade of grass and waiting patiently. When he pulls the blade out, it is covered with termites. David picks off the insects with his lips and eats them. You have observed David Greybeard using a tool to get his food!

Scientists such as Jane Goodall use questions to direct their investigations and plan what they need to observe. Jane Goodall wanted to know about chimpanzees. When she started her study, she watched everything the animals did. Because people did not know much about these animals, Jane Goodall needed to watch them for a long time to develop good questions to investigate. When she saw David Greybeard use the twig to get termites, she realized that chimpanzees' feeding was very interesting. She was able to ask a more direct question, "How do chimpanzees feed?" All of Jane Goodall's careful observations allowed her to answer her question about how chimpanzees feed.

collaborate: to work together.

Be a Scientist

Collaboration

Scientists **collaborate** to better understand the world around them. Like all scientists, they build on the work of others. They share their ideas by writing papers, or articles, in scientific magazines called journals. They also attend meetings with other scientists and present their work. Presentations and articles are ways biologists tell others what they have discovered. When other scientists improve or add to an idea, they write a paper about it and publish it for others to read. They always acknowledge where they got the idea. Journal articles and presentations are good ways for scientists to share their ideas.

Although Jane Goodall does not make observations of chimpanzees in the wild anymore, she continues to share her understanding of chimpanzees with others. She is particularly concerned about the future of chimpanzees. In 1977, she founded the Jane Goodall Institute. One of the goals of the Institute is to expand research programs on chimpanzees. Many new scientists now work in the Gombe area. They continue the work Jane Goodall started and build on her observations. One of the several questions they are presently investigating is *How do female chimpanzees interact with each other?* Another goal of the institute is to preserve all of Jane Goodall's journals in digital form. They then can be available to other scientists, as well as to the public.

Jane Goodall shares her understanding of chimpanzee feeding in Gombe. Other scientists will then discuss her information and use it to develop their own knowledge.

How Chimpanzees Feed

Jane Goodall and other ethologists have learned a lot about how chimpanzees feed. Chimpanzees are omnivores. This means they eat both plants and animals. Often, they eat fruits, leaves, and nuts from trees. At other times, they eat insects and small mammals. They choose what they eat based on what they can find. They move from place to place looking for food under rocks or in the bark of trees. Chimpanzees are foragers.

When Jane Goodall saw David Greybeard hunt termites, she was surprised to see that he used a tool to gather his food. Until that time, scientists thought humans were the only animals that used tools in that way. David Greybeard showed that other animals could learn to use tools.

David Greybeard and all the other chimpanzees could have just reached into the termite nests and grabbed the termites with their hands. Think about what might happen if they feed this way. The termite nest would get destroyed. By using a blade of grass or a twig to feed, the termite nest is not harmed. The termites can continue to live in it and make more termites.

At one time, scientists thought that chimpanzees only ate fruit, leaves, flowers, and roots. However, scientists now know that chimpanzees are omnivores.

If a chimpanzee ruins a nest in feeding, the supply of termites is lessened. Chimpanzees have found a way to adapt to their environment and protect it while they are feeding.

Chimpanzees are able to feed the way they do because of how their bodies work. They have hands similar to yours. They can pick up a single blade of grass or twig, hold it steadily and gently, guide it through a hole, and remove it carefully so the termites stay attached. Chimpanzees have been observed using other tools also. They crack nuts with rocks. Young chimpanzees learn to use tools by watching older chimpanzees.

Stop and Think

1. When Jane Goodall observed David Greybeard at the termite nest, she noticed something no one else had seen. Describe how you think Jane Goodall might have felt.

2. Describe the reasons why chimpanzees eat the way they do.

3. Jane Goodall watched the chimpanzees in the habitat where they lived. Why was it important to watch the animals in their own environment? What was difficult about watching the animals this way?

4. Jane Goodall made observations of the chimpanzees. Pretend you are Jane Goodall recording the observations. Record a descriptive observation of David Greybeard eating termites.

5. How does a forager obtain its food?

Reflect

Think about the observations and ideas presented in the previous section and what you read about in this section. Use an *Observing and Interpreting Animal Behavior* page, like the one shown, to record what you know about how chimpanzees eat. Include one behavior in each row of the chart. Then add what you know about what allows those behaviors and your interpretations of those behaviors in the other two columns. Collaborate with the members of your group to create one chart. Record all the ideas you have. Be prepared to share your observations with the class.

Observing and Interpreting Animal Behavior		
Name: _____ Date: _____		
Animal I am observing: _____		
Observations	What about the environment and animal allows that behavior?	Interpretations

Create Your Explanation

Name:_____ Date:_____

Use this page to explain the lesson of your recent investigations.

Write a brief summary of the results from your investigation. You will use this summary to help you write your Explanation.

Claim—a statement of what you understand or a conclusion that you have reached from an investigation or a set of investigations.

Evidence—data collected during investigations and trends in that data.

Science knowledge—knowledge about how things work. You may have learned this through reading, talking to an expert, discussion, or other experiences.

Write your Explanation using the *Claim*, *Evidence* and *Science knowledge*.

Explain

You now have learned a lot about chimpanzees and how they feed. You know that they are foragers and that they eat all different types of food. You also know that they can use tools to help them find food. Now you will write an explanation of how chimpanzees feed and what helps determine how they feed. Using a *Create Your Explanation* page, write an explanation of how chimpanzees feed and why they feed that way.

Your claim will combine the interpretations you recorded on your chart. It will state how chimpanzees feed. You have evidence from your observations showing that they feed this way, and you have learned a lot from your reading about chimpanzee feeding. The science knowledge you record should state the reasons chimpanzees feed the way they do.

Include in the science knowledge for your explanation what you now know about how a chimpanzee's body and habitat affect the way it feeds. Use all this information to help you develop your explanation of why chimpanzees feed the way they do. Your explanation will be a logical statement that connects your claim to your evidence and science knowledge. It will look something like this.

> Chimpanzees feed by [*tell how*] because [*give reasons from your science knowledge*]. We can tell they do it this way because we can see [*tell what you observed*].

Your explanation should match your claim, your science knowledge, and your evidence. If they don't match, revise your claim and your explanation until everything makes sense together.

Develop the best explanation you can using the evidence and science knowledge you have available. You will get a chance to write another one later in the Unit when you know more.

How do scientists answer big questions and solve big problems? Why do animals behave the way they do?				
What do we think we know?	What do we need to investigate?	What are we learning?	What is our evidence?	What does it mean for the challenge or question?

Update the *Project Board*

Earlier you began a *Project Board* focusing on the idea of creating an enclosure for animals. Now you have done some investigations and reading, and you know about how chimpanzees feed. You are now ready to record your knowledge about chimpanzee feeding on the *Project Board*.

You will focus on the third and fourth columns: *What are we learning?* and *What is our evidence?* When you record what you are learning in the third column, you will be answering some questions in the *What do we need to investigate?* column. But you cannot just write what you learned without providing evidence for your conclusions. Put your evidence in the fourth column. Some evidence will come from your data, and some will come from your reading.

You may use the text in this book to help you write about the science you have learned. However, make sure to put it into your own words. As your class works at the large *Project Board*, record the same information on your own *Project Board* page.

The *Project Board* is a great place to start discussions. You may find that you disagree with your classmates about what you have learned and the evidence for it. This is a part of what scientists do. These discussions help participants identify what they still don't understand well and what else they need to learn or investigate. Put any new questions your class develops in the *What do we need to investigate?* column.

What's the Point?

Feeding is a critical behavior for all animals. When animals feed, they eat food that gives them the nutrition they need. Different animals eat different foods.

Animals find food in different ways. Some animals forage. Chimpanzees are foragers. They gather plants and look for food in termite hills. Chimpanzees use tools to help them get termites out of their nests. Most other animals do not have hands that allow them to use tools the way chimpanzees do. Those animals have to get their food in different ways. The ability to use tools is one way chimpanzees are different from many other animals.

Because interpreting observations can be difficult, it is important for scientists to share their observations and ideas. This type of collaboration makes it possible for scientists to make sure their data is reliable. It also helps them build on each other's work. Building on the work of others allows scientists to answer questions they have developed but that they did not find answers to themselves. It also allows them to ask more detailed questions over time. For example, after Jane Goodall discovered that chimpanzees use tools, other scientists began to ask questions about whether other animals also use tools.

2.4 Investigate

How Do Bees Forage?

You have learned about the feeding behavior of chimpanzees. How do you think the feeding behavior of bees is similar to and different from the feeding behavior of chimpanzees? Interestingly, like chimpanzees, bees are foragers. Unlike chimpanzees, which are omnivores, bees are herbivores, which means they eat only plants. During the spring and summer, bees fly many kilometers away from their hive, searching for food. But how do bees know which flowers have the **nectar** and **pollen** they need? How do they choose which flowers to land on? Making these decisions will determine how successful a bee is at foraging.

Bees use nectar to make honey.

In this section, you are going to use model flowers, called flower cards, to simulate the foraging behavior of bees and the decisions that bees make when they are foraging. You will learn more about how bees forage and the special ways that bees are able to use their bodies to make foraging efficient.

Be a Scientist

Using Models and Simulations

A **model** is a representation of something in the world. A globe is one model that you know. The parts of the globe represent parts of Earth. Scientists use models to investigate things that are too difficult, too dangerous, too large, or too small to examine in real life. Models are also used to investigate events that would take too long or occur too quickly to observe in real life. The best models are designed so people can easily examine them to better understand something in the real world.

To learn from using a model, the model needs to be similar to the real world in ways that are important for what the scientist is investigating. Sometimes what you want to model is a situation or event. To do this, you create a model that includes the things that are part of an event and then use that to act out the event. This is called a **simulation**.

nectar: a sugary liquid produced by plants.

pollen: small, powdery grains that contain the male sex cells in seed plants.

model: a representation of something in the world.

simulation: use of a model to imitate, or act out, real-life situations.

A wind tunnel is used to simulate the effects of air moving over or around objects, such as airplanes or cars.

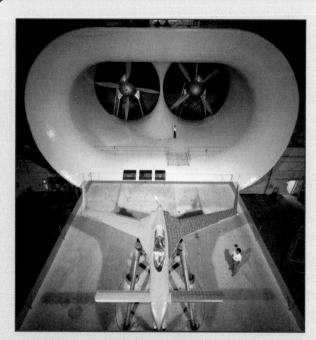

Simulations use a model to imitate, or act out, real-life situations in a way that is similar to real life but that allows you to examine what is happening without causing any harm or danger. Scientists use simulations when what they want to study is too big or too small, too fast or too slow, or too dangerous to investigate directly.

Models and simulations show the natural world as best they can, but there are always limits to how well a model can show what happens in nature. Nature is very complex, and sometimes models simplify nature too much. But to understand something complex, scientists design models that are simpler than the natural world. When you are working with models, it is always important to note how the model you are using is similar to and how it is different than the real world.

Now you will model how bees forage. Before you can understand the model and make comparisons between the model and the life in the hive, you need more information about bee life. Read the following carefully to better understand what bees do in the hive and how they are responsible for foraging for food.

colony (plural, colonies): a group of similar organisms living or growing together.

larvae (singular, larva): the newly hatched form of an insect (in this case, the bee).

Life in the Hive

Honeybees live in large groups, called **colonies**, inside a hive. There are three types of bees in the hive: the queen, the drones, and the workers. The queen bee lays all the eggs. The drones, the male bees, fertilize the eggs. Worker bees are all females. They build and guard the hive, take care of the **larvae**, produce the honey, clean the hive, feed the queen and drones, and collect nectar and pollen from flowers. They use the nectar to make honey to feed the colony. They use the pollen to feed the larvae.

Bees gather pollen from flowers. They are attracted to colorful patterned flowers. Bees recognize the color of the flowers they visit and whether or not that color or pattern flower has nectar or pollen. When a bee sees a color or pattern of flower that has nectar or pollen, it lands on that flower and collects its nectar or pollen. When it sees a color or pattern of flower with no nectar or pollen, it avoids that flower.

When a bee sees an unknown flower on a foraging trip, it has to make a decision about whether or not to land on it. Flying and collecting nectar and pollen take a lot of energy. If the unknown flower has nectar and pollen in it, the bee can collect its nectar and learn about another kind of flower it can collect nectar from. If the unknown flower does not have any nectar or pollen, landing on it uses up energy that could have been used for finding other food. But if a worker bee knows exactly where to find a food source, it saves a lot of energy and time. So sometimes it is useful to check a new kind of flower to see if it has nectar or pollen.

Each time they go out to collect nectar and pollen, they have to bring back enough nectar to feed themselves and help feed the queen and drones. The bees in a hive will starve unless the worker bees are good at finding nectar and pollen.

This means that bees have to find flowers with nectar. Landing on flowers without nectar is dangerous, because it can use up a bee's energy before it has a chance to collect enough nectar. But learning which flowers have nectar and which don't will help a bee become good at nectar collection.

ANIMALS IN ACTION

Stop and Think

Before you begin the investigation, use these questions to help you think about bees and how they forage.

1. Describe the environment of the bee and how you think that environment might affect how a bee will find food.

2. One reason chimpanzees are successful foragers is that they use their body in ways that help them forage. How do you think a bee's body can help it forage?

The Model

You are going to imagine you are a foraging bee. You have found a patch of flowers and need to collect as much nectar as possible while conserving energy. To do so, you must select the flowers with nectar and avoid the empty ones.

You and your partner will be given a set of 12 flower cards. Each card represents a different kind of flower. You will simulate the work of a foraging worker bee by selecting which of your flower cards the bee will visit.

You will want to select flowers with nectar while avoiding flowers without nectar. When you encounter an unknown flower, you will have to decide whether to check it for nectar. To do this, you will have to be able to identify which flowers have nectar. Read the box on the next page to know what information bees have about different flowers. This information will help you make good foraging decisions.

As you begin your foraging, pay careful attention to the differences between the flowers. Use what you know to land on flowers that contain nectar and avoid those that do not contain nectar. Remember that *every* visit to a flower with no nectar uses up energy without collecting more food.

Be a Scientist

Simulations Have Limitations

Simulations come close to how things occur in nature but also always have some limitations. In this simulation, you do not know if you have gathered nectar until your teacher tells you. A bee would know right away and might be able to visit another flower to gather more nectar. Also, as you will read later, the flower cards you use in this investigation are simpler than flowers are in nature.

> ## Simulation Rules
>
> What you (a bee) know about the flowers:
>
> - **Blue** flowers have nectar.
>
> - **Violet** flowers do not have nectar.
>
> - **Purple** and **pink** flowers are new. As a bee, you do not know if they have nectar until you land on them. You have to take your chances and choose which of these flowers to visit.

The Simulation

Remember that on each foraging trip the bee makes, it needs to bring back enough nectar to replenish its own energy and enough extra nectar and pollen to help feed the queen, the drones, and the larvae. In this simulation, you will forage for nectar by choosing cards from your deck to visit. You will not forage for pollen.

You will get 1 point for each flower you select that has nectar. You will lose 1/2 a point for each flower you choose that has no nectar. Your goal is to get at least 5 points—3 to represent the amount of food you (a bee) need to eat to get back the energy used in foraging, and 2 to contribute to the hive.

Procedure

1. Working with your partner, decide which of the 12 flower cards in your deck you will visit. Select between 5 and 9 cards. You may think you have a better chance of getting a full load of nectar if you choose more flowers. But remember that selecting the wrong flower results in subtracting points because of the extra energy a bee uses up.

2. Record your reasons for selecting each card. You and your partner may have some interesting discussions about which flowers to select. Perhaps you will even disagree about it. If you record what you talk about in these discussions, you will be able to participate in a discussion later about how bees might make their decisions.

3. After you have chosen your cards, look at the card numbers at the bottom right corners of the cards. Arrange your cards from lowest number to highest number. Put aside the cards you did not choose.

4. When everyone is finished, your teacher will tell you which of the flowers contain nectar and which ones do not. You will determine your foraging score based on this information.

Stop and Think

You and your partner made decisions about picking flowers based on a strategy you chose. Different students used different strategies and may have had different foraging scores. The higher the score, the more efficient your foraging was. Answer the following questions to prepare for a classroom discussion.

1. How did you decide which flower cards to choose?

2. How did you decide how many flower cards to choose?

3. What was your final foraging score? What could have helped you improve your foraging score?

4. Is there a way for a bee to successfully bring back enough nectar if it lands only on flowers that have nectar? Record how you think a bee could, or could not, do that.

5. How did the color of flowers help you as you looked for food? Record how you think the color helps the bees.

What's the Point?

Models and simulations help scientists better understand the world. Scientists use models and simulations to determine how things work. Models and simulations come close to how things occur in nature but also always have some limitations. In this simulation, you made decisions about what flowers to land on. Bees also need to make these decisions. Bees use the information they learn to help them make decisions. One way this simulation was different from nature is that the flowers you had were much simpler than flowers are in nature.

In this simulation, you had to make decisions with your group. The decision discussions may have been difficult, and you may not have agreed with your group. Scientists also have disagreements. Scientists sometimes disagree about scientific understandings. Usually, this is because they are working from different, or incomplete, evidence. In the next section, you will find more evidence to support your ideas.

2.5 Investigate

What Do Bees See that Helps Them Forage?

When you were simulating bee foraging, you probably wished you had more information about each of the flowers. Maybe you wished you knew what circles inside circles represented. Maybe there were other things you wanted to be able to see. In the next simulation, you are going to be able to see the flowers more like bees do.

The Model and Simulation

Just like last time, you are going to be a foraging bee. You and your partner will have 12 flower cards. Your goal will be to collect as much nectar as you can while avoiding landing on flowers that have no nectar.

One thing will be different this time. You will have a small flashlight that you can think of as a pair of "bee's eyes." In the last simulation, you used your human eyes to look for flowers. This time you will use the "bee's eyes" to look for flowers.

Procedure

1. Using the same flower cards as before, select between 5 and 9 cards to bring a full load of nectar back to the hive. Remember that selecting the wrong flower card results in subtracting points because of the extra energy the bee uses.

2. This time use the bee's eyes to help you make your foraging decisions. Select your flower cards and record the reasons for making your selections. Put the cards you did not select in a separate pile.

3. Listen as your teacher reads the scoring again. Keep track of your foraging efficiency.

Stop and Think

This time, you and your partner made decisions while using the bee's eyes. Think about how this was different from the first time you did the simulation. Answer these questions.

1. What information did the bee's eyes give you that you did not have before?

2. How did that information help you make decisions?

3. Look at the cards that you now know have nectar and the ones that do not. Try to find patterns on them that might help you identify next time which flowers have nectar. If you are having difficulty doing that, describe why.

4. What was your final foraging score? Why was your score different from the first time you did this simulation?

receptor cells: the cells that receive information from the world and send it to the brain.

ultraviolet (UV) light: a kind of light not visible to the human eye.

How Bees See

Bees and other insects have eyes adapted for the tasks they need to do. A bee's vision is important for foraging. When a bee lands on a flower that has food, it will remember the size, shape, and smell of that flower. It will then find other flowers with the same characteristics. Special patterns on flowers direct the bee to where the nectar is in the flower.

However, bee vision is very different from human vision. Bees see different things than humans do because the **receptor cells** in bees' eyes are different than those in human eyes. Humans see the range of light from red to violet. A bee's receptor cells are sensitive to green, blue, and **ultraviolet (UV) light**. UV light is not visible to the human eye, but it is visible to the eye of the bee.

Look at the pictures of geraniums in visible light and UV light. UV light reveals a secret. Some flowers have a dark spot in the center and lines leading from the petals to the center. These are visible only under UV light.

The same thing happened in your simulation. The flashlight you used is a UV flashlight. It allowed you to see markings on the flower cards not normally visible to you. The UV flashlight "changed" the UV markings into visible marks that you could see. A bee can see those markings without any special tools.

Bees' eyes allow them to see UV light.

| Geranium flower in visible light | The same flower in UV light |

Reflect

Use an *Observing and Interpreting Animal Behavior* page, like the one shown, to record what you know about how bees forage. Use your observations from the previous section and what you read in this section. Include one behavior in each row of the chart. Then add what you know about what allows those behaviors and your interpretations of those behaviors in the other two columns. Collaborate with the members of your group to create one chart. Record all the ideas you have. Be prepared to share your observations.

Observing and Interpreting Animal Behavior

Name: _____ Date: _____

Animal I am observing: _____

Observations	What about the environment and animal allows that behavior?	Interpretations

Explain

You have written several explanations. You are probably getting more used to developing explanations and supporting your claim with evidence and science knowledge. Develop an explanation about the foraging behavior of bees. Be sure to include information about bees' bodies and about the environment that might affect how they forage. Use information on your *Observing and Interpreting Animal Behavior* page to support your claim. Remember that your evidence and science knowledge can come from your experiences with the bee investigation and the science reading you have done. Reread the hamster example from *Learning Set 1* if you need more help with the different parts of the explanation.

Keep in mind that your explanation should be based on what you know now. You will have a chance to develop a more detailed explanation later as you learn more about bees and about what affects animal feeding.

Update the *Project Board*

You have now learned a lot about how bees forage. Add what you have learned to the third column of the *Project Board*. Make sure you add evidence from your investigations and reading in the fourth column. Then add questions to the second column. For example, you might wonder why bees see differently than people. As your class records their learning on the class *Project Board*, add the information to your own *Project Board* page.

What's the Point?

Foraging is one way animals feed. Foraging animals move from place to place to find their food. Some foragers, including bees and other insects, are herbivores. Some omnivores, like chimpanzees, are also foragers.

Bee foraging behavior is efficient because bees' bodies have adapted to finding pollen and nectar in flowers. Bees' eyes help make bees very efficient foragers. When bees look at flowers, they see them differently from the way other animals see them. Bees' eyes are tools that help them to be efficient in their environment.

electromagnetic radiation: a wave that travels through space and carries energy.

electromagnetic spectrum: the range of wavelengths emitted by the Sun.

visible spectrum: The part of the electromagnetic spectrum that can be seen by the human eye.

More to Learn

Light and Color Vision

The Electromagnetic Spectrum

Light from the Sun is a form of **electromagnetic radiation.** Electromagnetic radiation is a form of energy that can travel through space. Radio waves that you receive through your radio or TV are examples of electromagnetic radiation. This radiation travels as waves that differ in the amount of energy they carry. The energy is related to the properties of waves known as wavelengths, or frequencies. You will not read about the details of these properties here. For now, think of them as labels to identify the different radiations. It turns out that the different colors in sunlight can be distinguished by their wavelengths. The Sun emits radiation of many different wavelengths. The range of wavelengths emitted by the Sun is called the **electromagnetic spectrum**.

Human eyes can see only a small portion of the electromagnetic spectrum. This is called the **visible spectrum**. It covers the colors from red to violet. The red color has a longer wavelength than the blue and is less energetic. Next to the visible part of the spectrum are the infrared (IR) wavelengths, at lower energy, and the ultraviolet (UV) wavelengths, at higher energies.

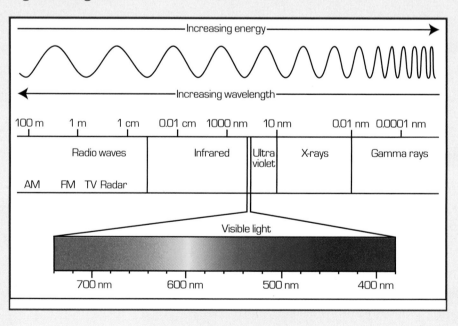

Humans cannot see IR and UV wavelengths, but some other animals can. For example, birds and some insects, like bees and some butterflies, can see UV light. Some animals, like snakes, can sense infrared light.

How Humans See Colors

Although small, the human eye is a remarkable organ. Basically, it works like a small camera. The light enters through a single transparent lens. The lens focuses the image on a layer that is sensitive to light called the **retina**. The retina lines the back portion of the inside of the eye. It contains two types of cells, **rods** and **cones**. The rods are the sensors that allow you to see in low light. The cones allow you to see colors.

The cones in the human retina are sensitive to red, green, and blue wavelengths of light. When the light strikes the retina, it generates an impulse that is transmitted to the brain by a nerve, called the optic nerve. By combining impulses received from the red, green, and blue cones, the human eye can see almost any color.

Objects appear to be a particular color because they reflect some of the wavelengths of the light they receive more than others. A red apple is red because it reflects light from the red end of the spectrum, and absorbs light from the blue end. When light reflected from an object strikes the retina, it stimulates the cones in such a way that you see the color.

retina: a membrane sensitive to light that lines the back portion of the inside of the eye.

rod: a type of cell found in the retina. Each rod is sensitive to low levels of light, but cannot see colors.

cone: a type of cell found in the retina. Each cone is sensitive to one color: red, green, or blue.

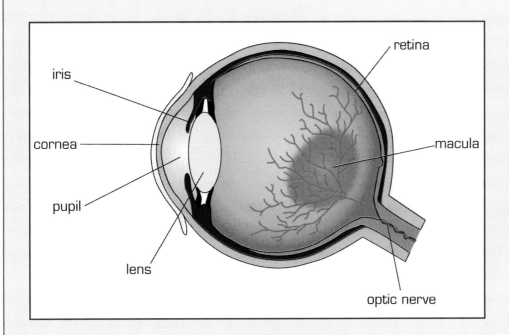

iris

cornea

pupil

lens

retina

macula

optic nerve

PBIS

2.6 Read

What Adaptations Do Bees Have that Affect Their Feeding Behavior?

reproduce: to produce offspring.

pollination: the transfer of pollen (male sex cells) from an anther to a stigma of a flower.

fertilize: in biology, to unite male and female sex cells to form a new organism.

In the last section, you learned about how bees' eyes work. Bees' eyes help bees forage because they have adapted to finding pollen and nectar in flowers. Bees and flowering plants also work together in other ways. In this section, you will read about how the relationship between bees and flowering plants helps both organisms survive.

How Bees and Flowers Help Each Other

Bees forage by looking for food in flowers. As they move from one flower to another, some of the pollen they find sticks to them and then falls off onto other flowers they visit. Flowering plants benefit from this process because bees move pollen from flower to flower. Combining pollen with the egg cell of a flowering plant makes it possible for flowers to develop seeds. Seeds are the way a flowering plant **reproduces** (produces more flowering plants).

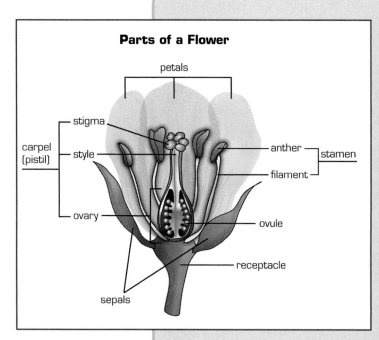

Parts of a Flower

petals

stigma

carpel (pistil)

style

anther

stamen

filament

ovary

ovule

receptacle

sepals

The diagram shows the parts of a flower. The production of *pollen* (male sex cells) takes place on an *anther*. The pollen is transferred to the *stigma*—a sticky surface at the top of the *pistil*. This process is called **pollination**. The male sex cells travel down the *style* to the *ovary*. In the ovary, the male sex cells **fertilize** the *ovules* (egg cells). The fertilized egg grows into an *embryo* (the early stage of an organism) that is surrounded by a supply of food and a protective layer. Together, the embryo, food, and protective layer form a seed. The seed can eventually produce another plant. You will learn more about the parts of a flower if you do the flower dissection that follows this section.

Most flowering plants rely on animals such as bees, other insects, birds, and bats to transfer pollen from the anther of one flower to the stigma of another flower. These animals are called **pollinators**. When a pollinator lands on a flower to eat the nectar, it brushes some of the pollen onto its body. When it lands on another flower, some of that pollen may drop onto that flower. For this reason, attracting pollinators is very important to flowering plants.

Flowering plants have adaptations that help them attract these animals. Adaptations help animals and plants survive in their habitat. Flowering plants have flowers with bright colors and fragrant smells to attract animals. The shape of a flower is also an adaptation. Sometimes the nectar of a flower is located deep inside the flower. The pollinator has to reach deep into the flower to gather the nectar. When it moves to another flower, it goes deep into that flower, too, and leaves some of the pollen behind—deep in the flower where it needs to be for reproduction.

Animals that are pollinators have also developed adaptations that let them reach the nectar inside certain flowers. Insects such as moths and butterflies have long, slender feeding tubes that can reach deep inside a flower.

Both pollinators and flowering plants benefit from their relationship with each other. The flowers provide the animals with food. The animals help to transfer pollen from one flowering plant to the next. Biologists call this type of relationship **mutualism**.

pollinator: an insect or other animal that carries pollen from one flower to another.

mutualism: a relationship between organisms of two different species in which each member benefits.

The buttercup has open petals. The petals have lines, like the ones you saw in your simulation, that direct insects to the sacs containing nectar in the center. Many different lightweight insects with short mouthparts are adapted to pollinate the flowers of these plants.

The sweet pea is adapted for pollination by insects like bumblebees. The nectar is inside a closed "cup" formed by a large petal and two smaller ones. The weight of a bumblebee landing on the large petal opens the cup. The bee can get the nectar inside. The pollen is transferred when the bee moves to another flower.

Wind-Pollinated Flowering Plants

Not all flowering plants are pollinated by animals. Some are wind pollinated. The flowers of plants that depend on the wind to transfer their pollen from one flower to the next are different from animal-pollinated flowers. The flowers of wind-pollinated plants are usually small and have no scent. Beacuse they do not have to attract animals, they have no nectar. They are also not very colorful. Most trees and grasses are pollinated by the wind.

Wind pollination is not as efficient as animal pollination. Therefore, wind-pollinated flowers need to produce large amounts of pollen to ensure that some of the pollen reaches the stigmas of other flowers. They must also rely on the weather. On rainy days and days without wind, very little pollen can get transferred from one flower to another.

The flowers on this tree hang down so that the pollen can be easily caught and blown by the wind.

The flowers of grasses are located at the top of the plant where the pollen is exposed to the wind.

Stop and Think

1. Adaptations are important for animals. Describe one adaptation of an insect and why it is important to that insect.

2. Sometimes the behavior of insects helps flowering plants. Without thinking about it, the bee moves the flower's pollen from one flower to another. Describe two ways that a flowering plant and an insect can have a mutualistic relationship.

Update the *Project Board*

You can now add what you have learned to the *What are we learning?* column of the *Project Board*. As you do this remember that you must support your learning with evidence. Put evidence from your reading and investigations in the *What is our evidence?* column.

What's the Point?

An animal's survival depends on its ability to gather and collect food. Adaptations are features of plants and animals that help them survive in their environment. Some adaptations help animals find and collect food. Sometimes adaptations of different organisms help the organisms work together to benefit each other. This is called mutualism.

For example, many insects rely on the nectar and pollen produced by flowering plants. At the same time, flowering plants rely on insects for reproduction. Flowers have adaptations, such as shape, color, and smell, that attract certain insects. The insects have adaptations, such as special vision or physical structures, that make it possible for them to get their food from the plant and help the flowering plants reproduce.

Butterflies feed on the nectar of a flower through long feeding tubes.

ANIMALS IN ACTION

A hummingbird is one animal that benefits from the nectar produced by flowering plants.

More to Learn

Flower Dissection

You have learned that flowers provide nectar and pollen that bees and other insects and animals need for food. Animals benefit from the flowers they feed on. However, the main purpose of flowers is to benefit the plants that produce them. Flowers are the reproductive organs of flowering plants, or **angiosperms**. The seeds that will develop into a new plant are produced in the flower.

In this section, you will perform a dissection of a flower to identify its parts and get to know what each part does. Based on the flower's structure, you will predict how it might be pollinated.

Daffodils

Gentian

Sweet pea

Anatomy of a Flower

Although flowers come in many sizes, colors, and shapes, they all have three major parts.

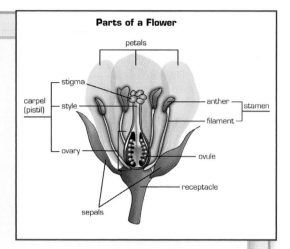

Parts of a Flower

petals

stigma

carpel (pistil)

style

ovary

sepals

anther

filament

stamen

ovule

receptacle

1. The **carpel**, or **pistil**, is the female part of the flower. It is normally shaped like a bowling pin. The carpel includes the **stigma**, on top, a **style**, shaped like a tube, and the **ovary**, at the bottom. The ovary contains the **ovules** that will develop into seeds after fertilization.

2. The **stamen** is the male part of the flower. It includes the **anthers**, where pollen is produced, and a filament that attaches each anther to the base of the flower. Each flower has several anthers arranged around the carpel.

3. The outer protective cover includes the petals, which are larger and colored, and the **sepals**, which are shorter and usually green. The colors of the petals serve to attract insects and birds that help to pollinate the flower. The petals form the **corolla**.

The flower parts are attached to the **receptacle** at the base of the flower. Not all flowers look the same. Some flowers have both female and male parts. Other flowers have only the female, or only the male, parts. The colors and shape of the petals and corolla also vary from flower to flower.

The basic function of the flower is reproduction. Pollination takes place when pollen from the anthers of a flower is delivered to the stigma of a flower. The stigma is covered with a sticky substance to which pollen grains attach. Once on the stigma, the pollen grain germinates and forms a **pollen tube** that grows down the style. The male sex cells, contained within the pollen, travel down the tube to the ovary. Fertilization takes place when the male sex cells unite with the ovules, or eggs. The fertilized ovule develops into a seed.

stigma: the top part of the carpel where the pollen is deposited.

style: in plants, the slender, tube-like part f the carpel.

ovary: in a plant, the enlarged part at the base of the carpel that contains ovules, or eggs.

ovule: a tiny, egg-like structure in flowering plants that develops into a seed after fertilization.

stamen: the male reproductive structure of a flower. It consists of the anthers, which produce pollen, and of a slender filament.

anthers: the parts on the stamen of a flower where pollen is produced.

sepals: green leaf-like parts of a flower found outside the petals.

corolla: the collection of the colored petals of a flower.

receptacle: the base of a flower to which the flower parts are attached.

pollen tube: the slender tube formed by pollen grains to reach and fertilize the ovules.

Materials

- newspapers, or paper towels

- a flower (azalea, lily, gladiolus, tulip, daffodil, geranium, snapdragon, or sweet pea)

- hand lens

- transparent tape

- scissors

- tweezers

- *Flower Dissection Observations page*

Procedure:
Observe the Inside of a Flower

Your teacher is going to provide you with a flower to dissect. Before you start your observations of what is inside your flower, it will be helpful for you to know a little more about the basic structure of flowers.

Read the section *Anatomy of a Flower* on the previous page to learn more about the main parts of flowers.

Working with other members of your group, spread the newspapers, or paper towels on the desk. Place the flower on the paper.

1. Observe the structure of the flower. Identify and count the sepals and petals. Keep in mind that the sepals and petals vary from flower to flower. The number, color, and shape of the sepals and petals in your flower may look different from the drawing in the *Anatomy of a Flower* section.

2. Use the tweezers to detach a sepal and a petal from the flower and tape them to the *Flower Dissection Observations* page.

3. Detach all petals from one side of your flower to expose its internal structure. Draw the inside of your flower on the space provided on the *Flower Dissection Observations* page.

4. Remove a stamen. Use your hand lens to inspect the stamen. Can you find the pollen? Use a piece of transparent tape to collect some pollen. Attach the transparent tape with the pollen to the space provided on the *Flower Dissection Observations* page.

Flower Dissection Observations

Name: _____ Date: _____

Use the space below to sketch, tape and describe the parts of your flower.

My flower has _____ sepals.
The color of the sepals is _____

> A sepal from my flower

My flower has _____ petals.
The color of the petals is _____

> A petal from my flower

Use the boxes below to sketch the inside of your flower and tape a sample of pollen.

> This is the inside of my flower.

> This is a sample of pollen from my flower.

PBIS-SE-AA-LS2_MTL-1BLM.doc

---

5. Use your hand lens to look at the end of the pistil. This is the stigma. Rub a stamen onto the stigma and observe it, again, with the hand lens. Record your observations on the *Flower Dissection Observations* page.

6. Use the scissors or your fingernails to open the ovary. Use the hand lens to locate and count the ovules inside. Sketch a diagram of the inside of the pistil on the *Flower Dissection Observations* page.

Communicate Your Findings

Investigation Expo

Once you complete your flower dissection, you will share your observations with the rest of your class. You will prepare a poster with the results of your flower dissection to present to your classmates. Answer the following questions to prepare for your classroom discussion.

- Identify and describe the female parts your flower has. Identify and describe the male parts your flower has.

- Based on the structure of your flower and what you know about how animals pollinate flowers, predict how it might spread its pollen around. Do you think the pollen is spread by the wind or insects? Why do you think so? Explain your answer with evidence from your observations of the flower's structure.

Prepare a poster with the results of your classroom discussion. Include the drawings and parts of your flower. Make sure you label the parts and describe their functions. Using what you know about the flower parts, describe how you think your flower is pollinated.

Present to your class your drawings of your flower parts and your observations of the stamen and pistil.

Share your prediction about how your flower is pollinated. Tell why you think it is pollinated that way, using evidence from your observations of the structure of your flower.

Listen carefully as others present their results and observations. Be prepared to discuss the evidence you and your classmates present to justify your ideas about how your flowers are pollinated.

2.7 Explore

What Are the Feeding Behaviors of Some Other Carnivores?

The animals you observed so far are foragers. They gather food by moving from place to place. Herbivores eat plants. Their bodies are adapted to eating and digesting plant material. Omnivores, such as chimpanzees, eat plants and animals. While chimpanzees are capable of hunting small prey and eating their meat, most of their diet is made up of fruit. Another group of animals eat only other animals. They are carnivores.

Some carnivores are predators that hunt, kill, and eat other animals. These carnivores must work hard for their food. The prey of some animals, such as lions or cheetahs, can move. This means the predator must be able to move faster. It must find and chase its prey in order to eat. Other predators wait until the prey comes to them. For example, alligators hide until their prey is near, and then they pounce. They use their powerful jaws to capture and consume their meal. Regardless of how predators get their food, their bodies must work like a well-designed machine, giving them the energy they need to overcome their prey. These animals have body parts that enable them to find, capture, kill, and eat other animals.

The cheetah is the perfect hunting machine with sharp eyesight, legs built for speed, teeth for tearing, and strong jaws to bring down its prey.

Conference

As an omnivore, you may have some experience eating meat. You may not hunt for your food. However, you have probably "hunted" for a meal at the grocery store and school cafeteria. Before you start this section, get together with your group and discuss what you already know about how a carnivore feeds. You can help get your discussion started by having one person in your group pretend to be a carnivore. As they pretend to "capture" and eat a piece of meat, you should watch their every move. Observe, and carefully record, the details as they go to get the food, put it in their mouth, and eat it.

When you are finished, use the following questions to guide your discussion.

- What actions and body parts were involved in

  a) obtaining the food?

  b) picking up the food?

  c) putting the food in the mouth?

  d) chewing and swallowing the food?

- What would be different if you were observing an animal in its natural environment hunting, chasing, capturing, and eating food?

- Describe any body parts or other characteristics that you think would make a predator a good hunter.

Predict

Use your group discussion to predict what characteristics and behaviors of carnivores help them meet their needs for obtaining food. Take into consideration that there are all kinds of carnivores—even a few plant species! What are some different strategies a predator might use to catch its next meal? Think about what forces it would take to overcome different kinds of prey.

Observe

As in other sections in this *Learning Set*, it is not possible to do direct field observations of a predator in action. In this case, it would be dangerous, as you could be viewed as prey by some animals! You may have seen some carnivores in the zoo. However, they do not have to hunt and capture their prey as animals in the wild do. The best way to observe the feeding behavior of carnivores is to watch a video. You will see three video clips. The first one shows a cheetah chasing and taking down a gazelle. The second one shows two lions taking down and eating a wildebeest. The third video clip shows an alligator laying in wait and then capturing a wildebeest.

Alligators use their powerful jaws to catch prey.

As before, when you watched the other videos, you will see each video twice. The first time, you will be watching it to come up with an observation plan. The second time, you will watch more carefully, using your plan, and then work with your group to identify all the different behaviors.

As you watch the first video clip, pay attention to how the cheetah's body is adapted for chasing and capturing prey. While watching the second clip, pay attention to how the lions' bodies and mouths are adapted for taking down, killing, and eating its prey. In the third clip, look carefully at what parts of the alligator are adapted for catching prey that is much larger than it is.

Plan

Discuss with your group what you observed about the way each animal feeds. Then identify what you need to observe more closely to be able to write a detailed description of how lions and alligators feed. Plan the way you will observe the videos the next time they are played so you will be able to gather more detail about the ways lions and alligators feed. What details will you pay special attention to in each video? Do you want to divide up the observations among your group? Should you each focus on different details? Remember that you want to be able to describe the following three things:

- how each animal feeds

- what is special about the body of each animal that allows it to feed that way, and

- what is special about the environment that affects the way the animal feeds.

Observe

Watch each video clip again, this time using your observation plan. Record what you are observing, paying special attention to your own assignment. Watch carefully to see what body parts help the cheetah chase its prey. Observe what body parts help the lions take down and eat their prey. Watch carefully to see how the alligator uses its great strength to overcome its prey.

Analyze Your Data

After observing the video, analyze the behavior of the animals as you have done for the chimpanzee and the bee. Share with each other what you observed. Record the behaviors you observed on sticky notes. Arrange the sticky notes in groups that allow you to interpret the behavior of each animal. Using a separate *Observing and Interpreting Animal Behavior* page for each animal, record the behaviors you observed, your interpretations of those behaviors, and what is special about each animal and its environment that allows it to feed the way it does.

You may ask your teacher to replay any parts of the videos that you did not understand or that you and your group members may have disagreements about.

Ask yourself how the bodies of the cheetah, the lions, and the alligator are different. Also, think about any similarities that you see.

Communicate Your Interpretations

Investigation Expo

In this *Investigation Expo*, some groups in the class will present their interpretations of the cheetah's behavior, others will present their interpretations of the lions' behavior, and some will present their interpretation of the alligator's behavior.

As you present, be clear about each of the following points:

- what you were observing
- your group's observation plan—what you were looking for and how you divided the work
- the behaviors you observed
- your ways of grouping the behaviors
- your interpretations of the behaviors, and
- what is special about your animal's body and the animal's environment that affects the way your animal feeds.

When you listen to other groups, be sure you understand their presentations. Ask questions if you need to know more. Notice observations and interpretations other groups have made that are different from yours. If you don't understand the reasons for the differences, or if you do not agree with their observations or interpretations, ask questions and state your disagreements. Any time you disagree, be prepared to present evidence that will allow you to support your claim. Remember to always be respectful, even when you are disagreeing with others.

Explain

You have just developed interpretations of how lions and alligators feed. Now it is time to develop an explanation of what affects how carnivores feed. Work with your group to do that. Your explanation should make clear why lions and alligators feed differently from each other. You probably predicted and observed that some carnivorous predators need to have good eyesight, strong legs for running and pouncing, sharp teeth, and strong jaws. Other predators may not be built for speed, but they have other adaptations, such as the ability to blend in with their environment until prey comes close enough to capture. Working with your group and using a *Create Your Explanation* page, develop an explanation of how a carnivore's body affects its particular feeding behaviors.

You might want to structure your explanation like this.

> Many carnivores feed by [*tell how*]. Others feed by [*tell how*].
> We can predict how a carnivore feeds by looking at its body parts.
> When a carnivore's body has [*tell what*], it feeds by [*tell how*].
> When its body is [*tell what*], it feeds by [*tell how*].

machine:
a device that makes work easier.

Incredible Predators

Predators have many adaptations that make them good hunters. They may have keen eyesight and hearing, the ability to blend in with their environment, specialized teeth for ripping and tearing flesh, and digestive systems that can break down meat efficiently.

Predators have to work to get their food. They must be able to chase down fast animals, capture prey that is sometimes larger than their own bodies, and kill and consume their food. To do this successfully, they must use force. Humans use **machines** to multiply force and make work easier.

In many ways, the bodies of animals are similar to machines. One example is the many **joints** found in an animal's body. A joint is the place where two bones come together. Some joints, such as those between the bones of the skull, are not movable at all. Others are slightly movable. Most of the joints found in the human body are **movable joints**. These joints allow a range of motion. Wherever there is a movable joint, you will find some type of **lever**.

joint: the place where two bones come together.

movable joint: joints that move freely.

lever: a simple machine made up of a rigid bar and fulcrum, or pivot point.

fulcrum: the point or support on which a lever turns.

effort force: the force applied to a lever.

load: the object being moved, or that resists the motion of the lever.

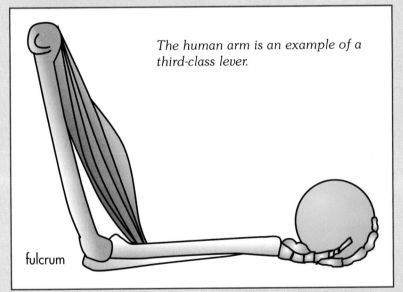

The human arm is an example of a third-class lever.

fulcrum

A lever is a simple machine made up of a rigid bar and fixed point. The fixed point that the bar rotates around is the **fulcrum**. Bones of the body act as levers with the joints acting as the fulcrum. The **effort force**, or the force applied to the lever, comes from the contraction of muscles. The object being moved, or that resists the motion of the lever, is called the **load**. Your arm is an example of a lever, with the joint at the elbow acting as the fulcrum.

Types of Levers

Levers work in different ways depending on where the fulcrum is positioned and where the force is applied. Differences in the position of the fulcrum, effort force, and load, result in three different types of levers.

ANIMALS IN ACTION

Types of Levers

First-class lever	Second-class lever	Third-class lever
A first-class lever has the fulcrum positioned between the effort and the load. A first-class lever changes the direction and multiplies an applied force. However, the effort force must be applied over a greater distance than the load.	*A second-class lever has the load positioned between the effort and the fulcrum. A second-class lever does not change the direction of the force, but it does multiply the effort force. However, the effort force must be applied over a greater distance than the load.*	*A third-class lever has the effort positioned between the fulcrum and the load. A third-class lever does not change the direction of the force. It also does not increase the distance of the effort force. However, the effort force moves over a shorter distance than the load.*
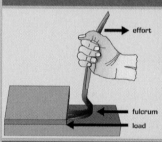	effort, load, fulcrum	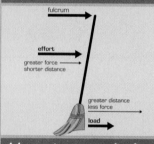 fulcrum, effort, greater force shorter distance, greater distance less force, load
A crowbar is an example of a first-class lever.	*A wheelbarrow is an example of a second-class lever.*	*A broom is an example of a third-class lever.*
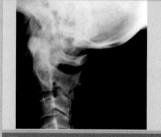	*There are no joints in the human body that are examples of second-class levers.*	
The joint between your skull and the top of your spine is an example of a first-class lever.		*Most of the movable joints in the body, including the elbow joint, are examples of third-class levers.*

Types of Movable Joints

There are several different types of movable joints in an animal's body. Movable joints give animals that hunt for food the ability to run, leap, lunge, and to clamp their prey in a deadly grip.

Ball-and-socket joints allow for circular motion. They are found in shoulder and hip joints. This type of joint gives a predator great flexibility of movement. Because the legs and arms are examples of third-class levers, force is not increased but the direction of the force is changed. This makes it possible for animals' limbs to provide greater motion.

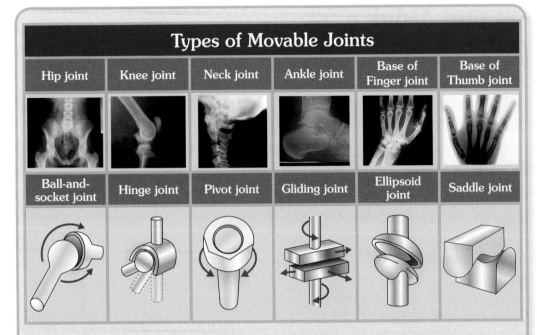

Types of Movable Joints

Hip joint	Knee joint	Neck joint	Ankle joint	Base of Finger joint	Base of Thumb joint
Ball-and-socket joint	Hinge joint	Pivot joint	Gliding joint	Ellipsoid joint	Saddle joint

Hinge joints permit a back-and-forth motion, similar to the movement of a gate swinging on hinges. They do not move from side to side. Your knee is an example of a hinge joint. Hinge joints are found in the jaws of many predators. Jaws are examples of two first-class levers with the point where they connect being the fulcrum. First-class levers multiply the force applied. This makes it possible for the animal to apply a lot of force with its jaws.

Pivot joints are characterized by one bone rotating around another. One bone is ring-shaped, and the other forms the pivot. This type of joint is found in the first two vertebrae of the neck. In birds of prey, such as owls, this type of joint lets them turn their heads and use their keen eyesight to spot prey.

Gliding joints occur where two bones with flattened or slightly curved surfaces come together. The two flattened surfaces can only slide past one another, giving this type of joint limited movement. Gliding joints are found in the wrists and ankles, as well as between the vertebrae that make up the spine. Having this type of joint in the backbone gives predators (as well as prey) an advantage. It gives four-legged animals the support they need while still allowing them to flex their backs. By flexing their backs, they gain power for movements such as running.

Ellipsoid and Saddle joints are commonly found in the bones of the hands and feet. Ellipsoid joints connect the fingers with the palms of your hands and toes to the feet. Saddle joints allow one bone to slide in two directions. The place where the thumb meets the wrist is a saddle joint. Both of these joints make hands and feet well adapted for climbing and grasping foods, which are behaviors typical of herbivores and omnivores.

Stop and Think

1. What advantages does a hinge joint give carnivores?

2. Why are most of the joints in a predator's body third-class levers? How does this benefit a predator?

3. Make a sketch of a carnivore that you are familiar with. Label each feature, or characteristic that makes the animal well suited for capturing and eating prey.

Revise Your Explanation

Based on what you have just read, revisit your *Observing and Interpreting Animal Behavior* pages. Add more detail to the middle column. If you need to, revise your interpretations.

Then, revisit your *Create Your Explanation* pages. You now have more science knowledge about what affects the feeding behavior of carnivores. Revise your claim if you need to, add your new science knowledge, and rewrite your explanation to add detail.

Share Your Explanation

Present your explanations to the class, and as a class, create your best explanation of what affects how carnivores feed.

Update the *Project Board*

Use the *Project Board* to record what you have learned about how carnivores feed. Be sure to link your learning statements with the evidence that supports them.

What's the Point?

Like all other organisms, carnivores are well-adapted for their feeding behaviors. You were able to observe and read about just a few different carnivores, but the world is full of many different meat eaters. Like lions and alligators, each has a different food preference, usually depending on its environment. Also, like lions and alligators, each is suited for capturing the particular prey they eat. Some have keen eyesight and sharp hearing, some have the ability to blend into their environment and lay in wait, and some have the ability to chase down fast prey.

The joints of all animals help the animal find and eat food. The joints of carnivores are particularly well-adapted for their feeding behaviors. Most joints are examples of third-class levers, meaning that the effort force is between the fulcrum (the joint) and the load. Muscles provide the effort force. Third-class levers increase distance over which the effort force is applied. This serves to increase the speed at which the animal can move. This is quite an advantage for any animal that must chase its dinner!

Learning Set 2
Back to the Big Challenge

You and your classmates have been developing an answer to the question, *What affects how animals feed?* Answering this question will help you complete the challenge of the Unit, to design an enclosure for an animal. Your enclosure should be designed so your animal can feed as naturally as possible. The enclosure should be as similar as possible to the animal's habitat.

To design your enclosure, you will need to think about some of the big ideas that affect how animals feed. Several big ideas were introduced in this *Learning Set*. You need to think about how those ideas are going to be built into your enclosure.

recommendation: a claim that suggests what to do in a described situation.

You are going to focus on one of the animals you learned about in this section. You will use observations, interpretations, and explanations of that animal's feeding behavior to help you develop a **recommendation** about designing an enclosure for that animal. You can also use information from the *Project Board*, especially the third and fourth columns, where you included information about each animal.

You will write a recommendation about how to provide the best environment for the animal so it can feed in the most natural way possible. Think about starting your recommendations with "If," "When," or "Because." For example, you might begin your recommendation by writing, "If the bees were going to feed in the most natural way possible, ..." Even better would be a recommendation of the form, "Because bees need to ..., ..." Then you need to complete the statement.

You will probably need to come up with more than one recommendation. Write as many recommendations as you need to write to make sure someone else will know everything that is important so that the animal's enclosure will allow your animal to feed naturally.

You will be sharing your recommendations with your class. You will want to convince others in the class that your recommendations are good ones. Because a recommendation is a type of claim, you will need to support your recommendation with evidence from your observations and reading. Some of this evidence may come from your previous explanations. Some will come from your observations and interpretations.

To prepare for presenting to the class, and so that you can be sure that your recommendations match the evidence you've collected and what science tells us, use a *Create Your Explanation* page for each recommendation your group makes. Your recommendation will be your claim. Add evidence and science knowledge that supports it. Then develop an explanation linking your recommendation to the evidence and science knowledge.

Create Your Explanation

Name:_____ Date:_____

Use this page to explain the lesson of your recent investigations.

Write a brief summary of the results from your investigation. You will use this summary to help you write your Explanation.

Claim – a statement of what you understand or a conclusion that you have reached from an investigation or a set of investigations.

Evidence – data collected during investigations and trends in that data.

Science knowledge – knowledge about how things work. You may have learned this through reading, talking to an expert, discussion, or other experiences.

Write your Explanation using the **Claim, Evidence,** and **Science knowledge**.

Be a Scientist

Making Recommendations

A recommendation is a type of claim that suggests to someone what to do when certain kinds of situations occur. It can have this form: When some situation occurs, do or try or expect something. For example, if you want to make a recommendation for crossing the street you might say the following:

When you have the right of way, expect that some cars will not have time to stop in time.

When you have the right of way, look both ways to make sure the traffic has stopped.

Recommendations might also begin with "if." For example, you might state another recommendation about crossing the street this way:

If you have the right of way, and the traffic has stopped, then you can cross the street.

You can also state a recommendation using a "because" statement.

Because some cars will not stop when the light turns red, make sure you look both ways carefully before you cross.

Communicate Your Solution

Solution Briefing

After you have developed your recommendations, you will communicate your recommendations to one another in a *Solution Briefing*. In a *Solution Briefing*, you present the solution you are developing in a way that will allow others to evaluate how well it achieves criteria and to make suggestions about how you might improve it. Your solution, in this case, consists of the recommendations you are developing. Before you start preparing, read more about *Solution Briefings* on the next page.

As you prepare for this briefing, make sure you revisit the criteria and constraints you identified in the beginning of the Unit. Use the following questions to plan your presentation:

- How is your enclosure addressing the feeding needs of your selected animal?

- How does the enclosure meet the criteria?

- How did the constraints affect your recommendations?

- What information did you use to help you make your recommendation?

- What other ideas did you think about along the way, and why did you not recommend them?

- What questions do you still have?

As you listen to your classmates' presentations, make sure you understand the answers to these questions. If you do not understand something, or if a group did not present something clearly enough, ask questions.

You can use the questions above as a guide. When you think something can be improved, make sure to contribute your ideas. Be careful to ask your questions and make your suggestions respectfully.

As you listen, record notes on a *Solution-Briefing Notes* page.

Solution-Briefing Notes

Name:_____ Date:_____

Design Iteration: _____

Design or group	How well it works	What I learned and useful ideas		
		Design ideas	Construction ideas	Science ideas

Plans for our next iteration

Be a Scientist

Introducing a *Solution Briefing*

A *Solution Briefing* is useful when you have made one or more attempts to solve a problem or achieve a challenge and need some advice. It gives you a chance to share what you have tried and learned. It also provides an opportunity for you to learn from others. You can ask advice of others about difficulties you are having.

Real-life designers present their design plans and solutions to others several times as they work on design projects. A team of designers sets up their design solution or design plan, and everyone gathers around. They make sure everyone can see. The design team presents its solution to everyone. The other designers ask questions and give helpful advice about ways to improve the design.

You will do the same thing. In a *Solution Briefing*, each team presents their solution in progress for others to see. Then teams take turns presenting to the class. Other classmates ask questions and offer helpful advice.

A *Solution Briefing* works best when everyone communicates well. Before you present your design or recommendation to the rest of the class, think about what might be important to share. What aspects of your solution should you present? What parts do you want to discuss with others? You need to be ready to justify to others what you decided to do and why.

When you are listening during a *Solution Briefing*, it is important to pay close attention. Look at each design or plan. Think about questions you would like to ask about each.

Each time you hold a briefing, you will take notes. You will fill out a *Solution-Briefing Notes* page as you listen to each group's presentation.

Update the *Project Board*

The last column on the *Project Board* helps you pull together all that you have learned during the Unit. In this column, you record partial answers to the *Big Question* or *Big Challenge*. Each investigation you do and each reading you complete is like a piece of a puzzle. You fit the pieces together to help you address the challenge. Each piece provides you with a critical factor that must be addressed to answer the *Big Question*.

The *Big Challenge* is to create an enclosure for an animal that makes it possible for the animal to feed or communicate as naturally as possible. The last column of the *Project Board* is the place to record your recommendations. Recording your recommendations on the *Project Board* as you go along will allow you to easily remember your recommendations when you return to the *Big Challenge* at the end of the Unit.

Learning Set 3

What Affects How Animals Communicate?

When you were a baby, you may have cried a lot. You cried to let someone know you wanted to eat, be held, or have your diaper changed. Your parents learned very quickly what your cries meant. By the time you were a year old, you were able to communicate in many new ways. You may have pulled on someone's sleeve, pointed to what you wanted, or even used a few words like "eat" or "more." By the time you were two, you could talk to others in two- or three-word sentences. These sentences communicated your wants and needs to your parents.

Communication is a behavior common to many animals. Baby animals often need to communicate to others their need for food or help. Communication can allow adult animals to live more comfortably. Once animals are able to communicate with each other, they can form groups and help each other find food and defend themselves from predators.

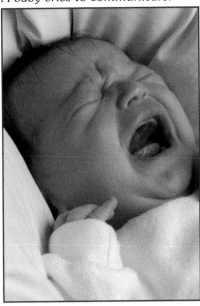

A baby cries to communicate.

To answer the *Big Question*: *How do scientists answer big questions and solve big problems?* you need to break it into smaller questions. In this *Learning Set*, you will investigate animal communication and answer the question *What affects how animals communicate?* You will read about and observe several different animals. You will examine how the form of the animal and the environment it lives in affects the ways it communicates.

A mother polar bear and her cub exchange greetings.

3.1 Understand the Question

Thinking about What Affects How Animals Communicate

The energy that food provides is a necessity of life. All animals find some way to feed. Another necessity of life is communication. Animals communicate when they need to affect the behavior of other animals. Human communication may include discussions about big and important ideas that affect the world. However, communication does not have to be complex. For most animals, communication is much more simple.

In this *Learning Set*, you will learn about different ways animals communicate. Some animals use sound, some use body characteristics like tails or coloring, some use gestures or facial expressions, and some use smell. More complex animals, including many mammals, use a combination of methods to communicate.

Animals communicate with other animals of the same species and sometimes with animals of different species. For example, dogs might bark at other dogs that enter their territory. Dogs also bark at other species of animals, like squirrels and people.

Communication is important to the survival of animals. Without the ability to communicate with others, animals would not be able to protect themselves, hunt for food, reproduce, find other animals in their group, or identify their territory.

A peacock communicates by spreading its tail feathers.

Wolves communicate with facial expressions.

Get Started

Humans communicate in many different ways. Their communication can be simple, like a smile or a frown, or more complicated, like a wink or a hand gesture. Humans also use language to communicate. Create a list of the different ways people communicate with one another, with and without using their voices, and with or without any tools. Record the kinds of situations and the reasons they communicate in each way. Use a chart like the one shown. One example is already recorded on the chart. Make your chart as detailed as possible. List at least five ways of communicating.

Observation Notes		
Ways of communicating	**Situations when people communicate this way**	**Why people communicate this way**
Jumping up and down and waving	The person you are communicating with is far away.	The jumping and waving can be seen from a distance. Sound might not carry far enough and it might be rude to yell.

Conference

Work with your small group to classify the different ways of communicating. Begin by reading to your group the different ways of communicating that you identified. On separate sticky notes, write each situation when people communicate that way. Then group the different ways of communicating under each situation. When you are done with your groupings, name each group. Make a list of your categories and the ways of communicating.

Communicate

Share your list with your class. Listen carefully as your classmates describe their lists. How are their categories different from or the same as your group's? Create a class list of communication categories.

Conference

You have considered communication in people. You have thought about different reasons why people communicate. Now think about questions you can ask about how other animals communicate.

Develop two questions that might help you understand what affects how animals communicate. Make sure the questions are not simply yes/no questions or ones you can answer with a single word or sentence.

When you write your questions, keep in mind that your questions should

- be interesting to you;

- require several resources to answer;

- relate to the *Big Question* or designing a new enclosure that will encourage the feeding or communication of an animal; and

- require collecting and using data.

When you have completed your two questions, meet with your small group. Share all the questions with each other. Carefully consider each question and decide if it meets the criteria for a good question. With your group, refine the questions that do not meet the criteria. Choose the two most interesting questions to share with the class. Give your teacher the rest of the questions so they might be used later.

Update the *Project Board*

The *Project Board* helps you to organize your ideas as you answer the *Big Question* and address the *Big Challenge*. You will now share your group's two questions with your class. Be prepared to justify why yours are good questions. Your teacher will add your questions to the *Project Board*. Throughout this *Learning Set*, you will work to answer some of these questions.

What's the Point?

Like feeding, communication is important for all animals. Some types of communication are simple. Language is a complicated form of communication. Humans use words to communicate with one another. They also use different methods like gestures and facial expressions. Reasons for communicating depend on the animals' needs for survival. Animals that live in groups need to be able to communicate with one another. Hunted animals or ones that need a lot of food communicate to protect their territory from invaders.

3.2 Investigate

How Do Humans Communicate?

Your class list of forms of communication helped you begin to see many different ways animals might communicate. In this section, you will think about what affects communication. You will explore two different ways humans communicate.

A mime communicates without words.

Predict

The goal for this activity is to solve a puzzle in two different ways. Some groups will use words to solve the puzzle and some will not use words. The groups will then analyze their observational data and share their conclusions with the class.

Before you begin solving the puzzle, predict which groups you think will be more successful at solving the puzzle: the "with–words" groups or the "without–words" groups. Begin by using your knowledge and past experiences to think about what instructions you might need to communicate to someone as you solve a puzzle together. How would you communicate those instructions, and what differences might there be between communicating with words and communicating without words? Then write your prediction as a statement that answers the following question:

> Will the "with-words" groups or the "without-words" groups be more successful at solving a puzzle together, and why?

You might start your sentence with, "I think the (*with-words/without-words*) groups will be more successful because, when solving a puzzle, it is helpful to..."

Be a Scientist

Predicting

When you write a prediction, you connect what you think you know about something, along with past experiences, to what you are learning now. Using all the information you have, you form an educated guess.

Procedure

For this investigation, you will work in groups of three. There will be two puzzle solvers and one observer in each group. The two puzzle solvers will solve the puzzle while the third group member carefully observes the puzzle solving and watches for successes and difficulties.

Some groups will follow the directions for "Solving a Puzzle with Words" and some for "Solving a Puzzle without Words." Read the directions very carefully because the two procedures are not the same. Notice the criteria and constraints that go with each set of instructions.

Materials
- **2 copies of a puzzle–one assembled, one unassembled**
- **paper**
- **pencils**

Solving the Puzzle without Words

1. The puzzle solvers sit facing each other.

2. One group member has an assembled puzzle, and one group member has an unassembled puzzle. The person with the unassembled puzzle should never see the assembled puzzle.

3. Using only facial expressions, hand gestures, or other means that do not involve speaking or writing, the two puzzle solvers should now solve the puzzle.

4. As the puzzle solvers are working, the observer should record observations about the work, paying attention to how the puzzle solving goes, watching the challenges and successes of solving the puzzle without words. The observer should also record how much time it took to solve the puzzle.

5. The time limit for solving the puzzle is 10 minutes.

Criteria	Constraints
Puzzle solvers should solve the puzzle together as quickly as they can.	Puzzle solvers have no more than 10 minutes to assemble the puzzle.
One puzzle solver can use an assembled puzzle for help.	The second puzzle solver cannot see the assembled puzzle.
Puzzle solvers can communicate only with facial expressions and hand gestures.	No words can be spoken or written.

Solving the Puzzle with Words

1. The puzzle solvers sit facing each other.

2. One group member has an assembled puzzle and one group member an unassembled puzzle. The person with the unassembled puzzle should never see the assembled puzzle.

3. Using spoken or written words, the two puzzle solvers will work together to solve the puzzle.

4. As the puzzle solvers are working, the observer should record observations about the work, paying attention to how the puzzle solving goes, watching the challenges and successes of solving the puzzle using words. The observer should also record how much time it took to solve the puzzle.

5. The time limit for solving the puzzle is 10 minutes.

Criteria	Constraints
Puzzle solvers should solve the puzzle together as quickly as they can.	Puzzle solvers have no more than 10 minutes to assemble the puzzle.
One puzzle solver can use an assembled puzzle for help.	The second puzzle solver cannot see the assembled puzzle.
Puzzle solvers can communicate with spoken and written words.	

Analyze Your Data

The first task in analyzing your data is sharing information between the observer and puzzle solver. Make a chart with three columns. Label them Observations, Interpretations and Conclusions. Using his or her recorded notes, the observer should share his or her observations with the puzzle solvers, and then add these observations to the first column of the chart. Add the observers' interpretations of what was happening during each observation to the second column. The puzzle solvers will know what behaviors the observer is referring to only if the observer provides enough detail about what the puzzle solvers were doing during each communication. **Ethnographers** (scientists who study people) call the exchange of information between groups a "member check." This is the first part of the member check.

Now, working with one observation at a time, the puzzle solvers should add more detail to help support the interpretations. This information should include why the puzzle solvers communicated the way they did. Include as much detail as you can and add this to the second column.. Observers will be able to understand the reasons why puzzle solvers communicated the way they did only if puzzle solvers are specific about why they communicated as they did. This is the second part of the member check.

Next, develop some conclusions about the different forms of communication your group members used. Identify the reason for each form of communication. Also, identify the difficulties of each form of communication. Record this information in the third column. Think about and record the kind of communication that would have been easier.

> **ethnographer:** scientist who studies people.

Communicate

Investigation Expo

Make a poster with your observations and interpretations. Your poster should include the following information:

- whether you were a "with-words" or "without-words" group

- your group's observations and interpretations

- the forms of communication your group members used, and what was easy and difficult to communicate using each one

- any conclusions your group reached about efficient communication

For this *Investigation Expo*, you will hang your posters on the wall. Everyone will have a chance to see all the posters. Then some groups will have a chance to present to the class.

As you look at the posters and listen to the presentations, think about the following questions:

- What similarities do you see in what each of the "with-words" groups observed?

- What differences do you see between what each of the "with-words" groups observed?

- What similarities do you see in what each of the "without-words" groups observed?

- What differences do you see between what each of the "without-words" groups observed?

- Which groups solved the puzzle more quickly—those who used words or those who did not?

- What conclusions did "with-words" groups develop?

- What conclusions did "without-words" groups develop?

- Which conclusions do you agree with, and which do you disagree with? Why?

Reflect

As you listened to other groups, you probably noticed the differences in puzzle-solving methods among the groups. You probably also noticed that some kinds of communication were easier with words and some were easier with gestures. With your group, discuss and record answers to the questions below. Be prepared to share your answers with the class.

1. Which form of communication was more effective in helping the solvers complete the puzzle? Why?

2. Describe the similarities and differences between the verbal and non-verbal (without words) forms of communication used in this investigation. What is non-verbal communication good for?

3. Some animals do not have the ability to communicate with words. How do you think they successfully communicate with one another?

4. There may be times when you need to communicate without words. Describe a time when that happened. How did you manage to communicate even without words?

Update the *Project Board*

While doing this investigation and discussing your findings, you may have thought of some new ideas or new questions. Use the *Project Board* to record your ideas about communicating with words and without words. Include in the second column of the *Project Board* any new investigation questions you might want to answer. Think about what you have learned and the evidence you have for your learning. Record your learning and the evidence on the *Project Board*. Think about how this learning will help you answer the *Big Question, How do scientists answer big questions and solve big problems?*

ANIMALS IN ACTION

What's the Point?

Communication is important for letting others know your wants and needs. You have seen through observations of humans that communication can happen without using spoken or written words. There may be times when it is necessary to communicate without words. There may have been times you experienced this yourself. However, humans do have the ability to use words. You were able to see in the investigation that this gives them a tremendous advantage when solving complex problems. All animals have the ability to communicate in some way. Sometimes constraints determine the form of communication.

3.3 Explore

How Do Bees Communicate and Why?

In the last *Learning Set*, you learned that bees forage for their food. Bees need to forage efficiently, because they use a lot of energy to collect their food. Bees have a unique way of communicating to other bees where they have found the food.

The bee in the center of the comb is performing a waggle dance that tells the other bees where food for the hive can be found. The dancing bee is blurry because it is moving quickly.

The Mystery of the Waggle Dance

The waggle dance is one of the most amazing behaviors found in the animal world. Karl von Frisch described the dance in the 1960s. Through careful observation, he noticed that bees would return to the hive with nectar and pollen and then shake themselves and turn around as the other bees watched. Karl von Frisch named this the waggle dance. He observed and recorded many bees engaged in this behavior and wondered why they did this.

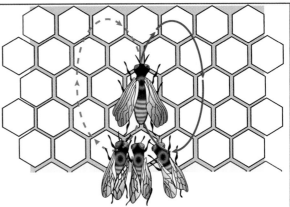

The bee in the center dances along the direction of the lines while the other bees watch.

ANIMALS IN ACTION

Observe

Think about what it might have been like to be von Frisch watching the bees and trying to develop an explanation of something other people had wondered about for centuries. You will watch a video of bees doing the waggle dance. To get an idea of what the bees are doing, watch the video once without taking notes. Think about what you will need to pay close attention to as you watch again. Remember that you will be trying to interpret the bees' behavior—what they are doing when they do the waggle dance, and why they are doing it.

Now watch the video again. This time, record your observations of the waggle dance. Record as much detail as you can. You might need to see the video several times before you feel ready to analyze what you saw.

Analyze Your Data

When you have completed your observations, work with your group to interpret the bees' behavior. Use an *Observing and Interpreting Animal Behavior* page to help you.

Observing and Interpreting Animal Behavior

Name: _____ Date: _____
Animal I am observing: _____

Observations	What about the environment and animal allows that behavior?	Interpretations

Decide which observations of behavior to record in the Behavior column. Include in the second column what you know about bees' bodies and their environment that might affect that behavior. Then develop some interpretations. Why do you think the bees were doing what they were doing? What does it look like the bees are trying to communicate? What messages do you think they are giving to other bees?

You may not know about bees yet, so your interpretations will be educated guesses. After you read more about bee communication, you will revisit your *Observing and Interpreting Animal Behavior* page and add more detail.

Why Do Bees Communicate?

Bees are very social animals. They live in large groups in one hive. Their survival depends on cooperation in the hive. Each bee has a job to do. The job of the queen bee and drones (male bees) is to make more bees by reproducing. The rest of the bees are female worker bees. They do not reproduce. Recently hatched worker bees clean the hive and care for the queen's offspring. As the worker bees mature, they build parts of the hive. Later, they guard the hive from predators. The last job of a worker bee is to leave the hive and forage for food.

If a beehive is to be successful, the bees need to make more bees. So they must have food. You already know that the worker bees have responsibility for getting food and bringing it back to the hive. And you know that each of them discovers new flowers with nectar or pollen as they forage. Remember that the worker bees need to be very successful at this task or they will not be able to supply enough food to the hive. It seems only natural, then, that they should have a way to communicate with the other bees. But remember that, up until recently, scientists thought the bees were dancing only to get attention.

Worker bees feeding the queen bee.

Karl von Frisch thought there must be other reasons for the dance. It seemed too complicated to be just for getting attention, and he had seen bees become more efficient at locating food. He observed honeybees and performed experiments to better understand the purpose of the dance. From his observations, he concluded that bees waggle to show other bees in the colony the location and distance of food from the hive. Other worker bees then use the information from the dance to quickly locate food. This saves a tremendous amount of energy, since the worker bees do not have to fly around randomly, looking for food sources. This discovery was very important. It earned Karl von Frisch a Nobel Prize in 1973.

Even though Karl von Frisch won an important prize for his discovery, some scientists still disagree with his idea. One scientist, Adrian Wenner, started out thinking the waggle dance was a form of bee communication but now thinks bees communicate through scents. Adrian Wenner and others think bees could be marking the flowers with scents from their bodies. Then the other bees could follow the odors to find the food.

Worker bee with attached radio antenna.

Scientists have debated different theories and worked to explain why bees did the waggle dance. The problem they faced was that they could not follow individual bees after they left the hive.

New technologies are making it possible for scientists to make more detailed observations of bee behavior. In 2005, a group of British scientists glued tiny radar antennae to worker bees. The radar equipment made it possible to follow the path of individual worker bees. Scientists could also monitor worker bee movement over time. The researchers concluded that the waggle dance is indeed the way bees share information about food with one another. These scientists concluded that Karl von Frisch was right. Their interpretation of the observations leads them to agree with him that the dance shows other worker bees the location of a food source and its distance from the hive.

The controversy is not settled though. Many bee scientists think the evidence supports the waggle-dance conclusion. But Wenner and other scientists question the data and evidence presented in the radar studies. They still think the bees communicate by leaving an odor on the flowers and having other bees find the odor. Scientists are identifying new ways to investigate bee behavior so they can know for sure.

Stop and Think

1. Scientists are still debating the use of the waggle dance. Why do you think bees use the waggle dance? What makes the waggle dance an effective way for bees to communicate?

2. What would happen to human communication if the only way to communicate was through a human "waggle dance?" What things would we have a lot of trouble communicating? How would we be able to do science?

3. Karl von Frisch first wrote about the waggle dance in 1927. He won the Nobel Prize in 1973, and scientists continue to debate the use of the waggle dance. Everyone agrees that bees move in certain ways, but they still don't agree about why bees do the dance. Why do you think the debate has continued for so long?

Update the *Project Board*

As you observed the waggle dance and interpreted the bees' behavior with your group, you may have thought of new questions you want to ask about animal behavior. You also learned about how scientists interpret the waggle dance. Add your questions and what you learned to the *Project Board*. Make sure you add evidence to support any new science you learned. Also, think about how what you just learned can help you answer the *Big Question: How do scientists answer big questions and solve big problems?* As your teacher records this information on the big *Project Board*, add the information to your own *Project Board* page.

What's the Point?

One of the most important tools for animal survival is the ability to communicate. Animals can exchange information about food and water sources, possible dangers, and mating. Scientists think bees communicate using a waggle dance. Other scientists think bees use scent markers.

Although bees do not have bodies that allow them to communicate through gestures, they have developed an interesting dance that only they fully understand and use to communicate with other bees.

Scientists debate different scientific ideas. They use scientific methods to investigate their ideas. They analyze the data and generate conclusions supported by the data. Sometimes, theories are so complicated and difficult to investigate that debate over them lasts for many years. The waggle dance is one theory that has been debated for decades. Scientists continue to question the scientific methods used to study the waggle dance.

Bees use the waggle dance to communicate the location of food to other bees in the hive.

3.4 Explore

What Affects How Elephants Communicate?

You found when you were solving the puzzle that communicating with words is more effective in many ways than communicating without words. Spoken words can be used when you can see the person you are communicating with and when you cannot see the person you are trying to communicate with. For example, if you were in a large store, you could call for your parents even if you could not see them. But often, verbal communication works better when someone is making eye contact with the person they are speaking to. Then facial expressions and body language help send the message.

Humans and bees use different methods to communicate. Communication methods are affected by an animal's body structure. Humans use different parts of their bodies to communicate than bees do. Humans can use their hands, faces, and entire bodies. Bees do not have all these body parts to use. Their bodies do not allow for the range of communication that humans have. Some of the ways animals communicate are not visible. Bees communicate through odors and dances. Odors are not visible, but they still provide information for communication.

An elephant can communicate using its eyes. An elephant's eyes can show that it is aggressive or is paying attention.

Elephants communicate in some ways that are similar to bees and humans. They also communicate in ways that are different from these two animals. As you learn more about elephant communication, you should pay attention to three different things that affect how elephants communicate. First, think about how an elephant's body affects how it communicates. Next, pay attention to ways elephants might communicate that cannot be seen, such as using odor. Finally, think about how an elephant's environment affects the way it communicates.

savanna: a grassland area with scattered trees, usually found in tropical or subtropical regions.

Predict

The picture shows some African elephants as they walk in a group. These elephants live on the **savanna**, a large, mostly grassy area of Africa. Just like people and bees, elephants live in groups. They need to communicate with other members of the group. Predict how these animals communicate with one another. What tools do you think elephants use to help them communicate? Think about what you learned about how humans communicate and the tools humans use. Identify any similarities and differences between humans and elephants. You can use these questions to help you think about elephant communication.

- What are some reasons animals need to communicate?

- What are some different ways animals communicate with one another?

- Think about the way an elephant's body works. How do you think elephants use their bodies to communicate with one another?

- What type of elephant sounds do you think an elephant uses to communicate with another elephant?

A group of African elephants walking on the savanna.

Observe Elephant Communication

It would be difficult for your class to make field observations in the African savanna. Although a safari to Africa to observe elephants would be exciting, it is probably not possible. You may have observed elephants communicating at a zoo. However, the zoo is not like the savanna. Once again, the best way to observe elephant communication is to watch a video.

As when you watched the other videos, you will see the video twice. The first time, you will be watching it to come up with an observation plan. The second time, you will watch more carefully, using your plan, and then work with your group to identify all the different forms of communication.

Plan

Discuss with your group what you observed about the way the elephants communicate. Then identify what you need to observe more closely to be able to write a detailed description of how they communicate. Plan the way you will observe the video the next time so that you can make detailed observations. What details will you pay special attention to in each video? Do you want to divide up the observations among your group? Should you each focus on different details?

Remember that you want to be able to describe three things.

- how elephants communicate
- what is special about the elephant's body that allows it to communicate that way
- what is special about the environment that affects the way elephants communicate

Observe

Remember to follow the plan your group decided on. Record your observations so you will be able to share them with your group and your class. Once you have your plan written and all the members of your group understand the plan, you will watch the video again. Remember to take notes about the elephants' behavior and habitat. Pay attention to the elephants' bodies. Think about how their body structure helps them to be better at communicating. Think also about how an elephant's habitat affects how it communicates.

Analyze Your Data

The video you watched showed elephants communicating. How did you know the elephants were communicating? You may have predicted and observed that elephants make different sounds. These sounds communicate a variety of things to other elephants. This is one way other mammals, such as people, communicate.

Meet with your group to share your observations. Each member of the group should have a chance to share all their observations. Remember that, depending on your observation plan, you may not have been watching what the other group members were watching, and they may have a lot of information to share with you. Listen carefully for observations you might not have seen. Create a group list of observations that everyone agrees on. Make another list of observations that were not agreed on. Save the observations you could not agree on for the *Investigation Expo.*

Using the same procedure you used in *Learning Set 1*, create a sticky note for each observation on which you all agreed. Organize your observations into groups based on the type of communication that was shown. For example, how an elephant uses body movements to communicate might be a group of observations. If you think that some forms of communication might fall into two groups, you can list those twice.

After identifying the types of communication you observed, it will be time to think about what your observations mean. Think about when elephants make different sounds and what the different sounds might mean. Think about the different ways elephants use their bodies to send information to each other and what those body movements might mean. These interpretations will help you as you begin to develop explanations.

Using an *Observing and Interpreting Animal Behavior* page, record what you know about how elephants communicate. Record your observations in the left column. Include one behavior in each row of the chart. Then add what you know about what allows these behaviors in the middle column and your interpretations of these behaviors in the right column. Collaborate with the members of your group to create one chart for your group. Be prepared to share your observations and interpretations with the class.

Observing and Interpreting Animal Behavior

Name: _____ Date: _____

Animal I am observing: _____

Observations	What about the environment and animal allows that behavior?	Interpretations

ANIMALS IN ACTION

Communicate

Investigation Expo

You will share your observations and data analysis in an *Investigation Expo*. In this *Expo*, each group presents their poster to the class. In your presentation, you will share what you have seen with the class. Remember that each group created their own plan for making their observations and analyzing their data. Groups' observations may have been affected by their plan, and the analyses of other groups may be different from yours.

For your presentation, create a poster that describes each of the following:

- the questions you were trying to answer in your observations
- your observation procedure and how it helped you make observations
- your analysis of your observations and how confident you are about the analysis

At the bottom of your poster, make a list of the observations your group did not agree on.

Before the presentations, read the questions below. Plan your group's presentation so that you answer all of the questions. Then as each group is presenting their poster, listen for the answers. If a group does not answer all these questions, ask them to help you understand what they found out. When you ask the questions, focus on better understanding the observations the group made.

- What was the group trying to find out?
- What procedure did they use to collect their data?
- Were they able to make clear observations that were detailed?
- How did they group their observations? What did their groupings allow you to see?
- What conclusions do their results suggest?
- Do you trust their observations? Why or why not?

Explain

You have written several explanations in this Unit. You are probably getting more used to developing them. Using your observations and interpretations, create an initial explanation of what affects elephant communication. Your explanation of the behavior must bring together the evidence (observations), the interpretations your class agreed on, and your science knowledge. In this case, your science knowledge may come from something you read, something you saw in the video, or from past experiences. Using a *Create Your Explanation* page, record your explanation of why elephants communicate as they do. Remember that the best explanations help others understand what makes a claim accurate. After you read more about elephant communication in the next section, you will have an opportunity to revise your explanation.

Communicate

Share Your Explanation

Share your explanation with the class. As each group shares their explanation, pay special attention to how the other groups have supported their claims with science knowledge. Ask questions or make suggestions if you think a group's claim is not as accurate as it should be or if the group has not supported their claim well enough with observations and science knowledge.

What's the Point?

Humans and bees use different methods to communicate. Communication methods are affected by an animal's body structure. Some animals, like humans, have more body parts to use for communication than other animals. Some methods of communication, such as odors, are not visible.

Elephants communicate in some ways that are similar to bees and humans. They also communicate in ways that are different from these two animals. Elephants use body movements, sounds, touch, and odors for communication. Their environment also affects the way elephants communicate.

In science, observations are usually done before interpretations. First, scientists observe without making interpretations. Then, they use their observations and other information to make interpretations. Lastly, using the interpretations of their observations and all the evidence they have gathered, scientists develop explanations.

3.5 Read

How Are Elephants Adapted for Communication?

Sound is a very important form of communication for elephants. One truly amazing thing about elephants is the number of different sounds they can produce. Researchers have found that elephants can produce at least 25 different sounds. These sounds include rumbles, trumpets, snorts, and screams. Using these sounds, elephants tell their group members about how and where they are. They can send information about food and safety, as well as about playfulness and pleasure.

Making Sounds

By changing the length of its trunk, an elephant is able to change the sound it makes.

The elephant has a huge head and body. This means that sounds an elephant makes can vibrate and echo through its body. Every part of the elephant helps it make sounds. The way an elephant holds its head, trunk, or throat can change how a sound is made.

Types of Sounds

One common way elephants communicate is called a rumble. Rumbles are very low pitch sounds. Because they are low pitch sounds, the rumbles can be heard from very far away. Elephant groups spread over large distances to find food. Also, the savanna where the elephants live is flat and mostly treeless. Low sounds travel very well in this environment where there are few objects for the sound to bounce off. So the rumbles are a perfect sound for elephants that need to communicate over long distances.

Until recently, scientists did not know that elephants made these low rumbles. That is because the rumbles are too low for people to hear. Scientists observed elephants and how they seemed to communicate with one another. They thought the only way elephants could be communicating was through sound, but they were not hearing anything. These observations made scientists think they needed to listen to the elephants more carefully. Scientists needed tools that would allow them to listen to lower sounds.

By using microphones that can pick up sounds below human hearing, they were able to hear the rumbles of elephants.

Using these tools, scientists found out that elephants also make other noises. Elephants have a tremendous range of sounds. They can make low sounds and high sounds. They can make quiet sounds as well as loud sounds.

Making a wide range of sounds does not really help an elephant if it cannot hear the sounds. An elephant's hearing is also amazing. Not only can elephants hear rumbles over long distances (10 km, 6.2 mi), they can also determine the direction of the sound. Just by listening, elephants are able to know where their group members are, even if the other group members are far away.

An elephant has amazing hearing that can pick up sound over great distances.

What Are Some Other Ways Elephants Communicate?

Think back to when you were working on the puzzle. Some students used only gestures to help their partner assemble the puzzle. These groups found that they could be effective with gestures, but that it required being able to see each other. Elephants also use different parts of their bodies to communicate visually. In a threatening situation, an elephant will stand very tall with its head held high. Sometimes it will even stand on a log or rock to appear taller. When an elephant flaps its ears rapidly, it means that it is excited. A large male elephant may kneel as an invitation to a smaller elephant to play. Some elephants have even been observed smiling.

Touching is another way elephants communicate with their bodies. Elephants use their trunks to greet one another. Female elephants will entwine their trunks in a way that is similar to a human handshake. They also use their trunks to caress one another. When mother elephants brush their **calves** with their trunks, they seem to be comforting them. Elephants can also be aggressive. They can use their trunks to uproot and throw things, or use their tusks to push one another.

One way elephants communicate is not easily observable. Scientists know about chemical communication from many different types of animals, including humans. Elephants have a keen sense of smell. Other than sound,

calf (plural, calves): the young of certain kinds of animals, such as an elephant or cow.

An elephant calf closely follows its mother. To comfort the calf, the mother may touch it with her trunk.

elephants probably use chemical communication more than any other type of communication. Elephants are often seen using the tips of their trunks to follow trails marked by urine or other waste material.

Stop and Think

1. What are some ways elephants communicate with each other?

2. People communicate about a variety of things. What are some things elephants must communicate about?

3. How does an elephant's environment affect how it communicates?

Revise Your Explanation

You have just read more about how elephants communicate. With your group, look back at your *Observing and Interpreting Animal Behavior* pages. Look at your original interpretations of elephant communication. Now, with the new science knowledge from what you have read, reinterpret the behavior you saw in the video.

Go back to your explanation on your *Create Your Explanation* page. First, add the science you just learned to the science knowledge box. Then, check to make sure your claim is still accurate. If your claim does not match the science you have read, you should revise it. Next, support your claim with the science knowledge you just learned.

Rewrite your explanation to make it more complete. Remember that an explanation is a statement that connects a claim to evidence and science knowledge in a logical way. Write your explanation so that it tells why your claim is accurate. Be sure your explanation matches the science you just read. Make sure your claim now matches what you have learned. If it does not, revise your explanation. Use the information from your reading about elephant communication to support your revised explanation. You might need to write an explanation that has a few sentences rather than just one long sentence. This is fine. The goal is to to tie everything together and help others understand why elephants communicate the way they do.

Communicate

Share Your Explanation

Share your new explanation with the class. When you share your explanation, tell the class what makes this revised explanation better than your earlier one. As each group shares their explanation, pay special attention to how the other groups have supported their claims with science knowledge. Ask questions or make suggestions if you think a group's claim is not as accurate as it could be or if the group has not supported their claim well enough with observations and science knowledge.

As a class, create your best explanation of what affects how elephants communicate.

Update the *Project Board*

As you read more about how elephants communicate, you may have thought of new questions you want to ask. Add what you learned from your reading to the third column of the *Project Board*. Make sure you add evidence to support any new science you learned. Also, think about how what you are learning will help you answer the *Big Question, How do scientists answer big questions and solve big problems?* As your teacher records this information on the big *Project Board*, add the information to your own *Project Board* page.

What's the Point?

Animals communicate in a variety of ways. An elephant uses gestures, sound, and smell to communicate with others. This is similar to how humans and other mammals communicate. How each type of animal communicates depends on its physical characteristics. The elephant uses its trunk, head and body size, and ears to help with communication. Also, the elephant's environment affects how it communicates. The sounds an elephant makes are tailored to meet the needs of the savanna where it lives. Animals, including humans and elephants, communicate to keep track of family members, to alert others to danger, to identify a food source, to play, and to take care of their young.

More to Learn

What Is Sound?

Sound is one way animals can communicate with each other. You have read about how elephants communicate using sound. But what exactly is sound?

Matter is made of small particles, called molecules. Sound is created by the back-and-forth motion, or **vibration**, of the molecules in matter. Because sound is the movement of molecules, sound can move through gases, liquids, and solids. Sound cannot travel where there are no molecules to move.

Sound as Waves

When you make a sound with your voice, your vocal chords vibrate very quickly. The vibration causes molecules in the surrounding air to be pushed together in a pattern. Between each of the places where the air molecules are pushed together is a part where the air molecules are spread out. This pattern is repeated over and over again very rapidly. Scientists call this movement of molecules a **sound wave**.

As the vibration in your vocal chords begins, the air nearest your vocal chords is pushed together. This group of air molecules that has been pushed together is called a **compression**. Next, the vocal chords move back, giving the air molecules room to spread apart. The space where molecules are spread apart is called a **rarefaction**.

The diagram on the next page shows a vibrating tuning fork. All sound waves happen in the same way, regardless of how they are made. They all have compressions and rarefactions. The vibrations that create sound are repeated very rapidly, so rapidly that you cannot see them. The sound waves then move through the matter that surrounds the source of the sound. Often, the matter is air, but sound can also move through liquids and solids.

You have probably noticed that the farther you are from the source of a sound, the softer the sound is. When you are very close to the source of a loud sound, the sound may even hurt your ears. But for any sound, no matter how loud, from very far away, you cannot hear it at all. This is because sound waves change and lose strength as they move away from their source. For example, as you speak, the sound waves you make move away from you. As they move away, each wave spreads out, which decreases the strength of the wave at any point.

vibration: the back-and-forth movement of molecules.

sound wave: the movement of molecules in a pattern, repeated over and over again, very rapidly.

compression: part of a sound wave where molecules are pushed together.

rarefaction: part of a sound wave where molecules are spread apart.

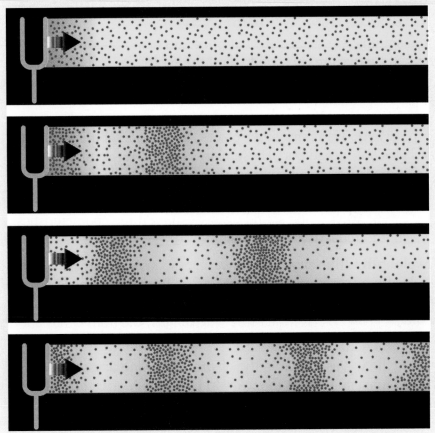

Air molecules are pushed together as the tuning fork arms vibrate quickly. The compressions move away from the tuning fork as sound waves.

Also, when sound waves run into another object, they either bounce off it or are absorbed by it. When waves bounce off an object, some of the wave strength passes into the object. The remaining wave is weaker.

When an object absorbs waves, the sound cannot move any farther. So sounds in wide-open spaces can move farther than sounds in closed areas, or crowded areas, such as areas with many trees. The objects in closed-in or crowded areas absorb sound waves, which decreases the amount of sound traveling through them.

Sound waves travel at different speeds in different kinds of matter. Molecules in gases are very spread out. Molecules in liquids are packed together more tightly. Molecules in solids are packed together even more tightly than in liquids. Because molecules in gases are more spread out than those in liquids, sound moves more slowly in gas than in liquid. Because molecules in solids are packed even more closely together, sound waves move fastest through solids.

wavelength: measured from one compression to another compression, or from one rarefaction to another rarefaction.

frequency (of waves): the number of waves that pass a point in a second.

pitch: a measure of the frequency of the vibration of the source of a sound.

Characteristics of Sound Waves

The diagrams shown on the next page can help you imagine sound waves. If there were no sound, all of the molecules would spread out evenly. A vibrating object pushes molecules together in some places. You can see that on the diagrams there are places, shown as dark areas, where the molecules are pushed together. These places are the compressions. There are other places, which are lighter, where the molecules are spread out. These places are the rarefactions. A single wave is made up of one compression and one rarefaction.

Scientists measure certain characteristics of waves. All waves—light waves, radio waves, and sound waves—share three characteristics: wavelength, frequency, and amplitude.

Wavelength is the length of the wave. Scientists determine wavelength by measuring individual waves from one compression to another compression, or from one rarefaction to another rarefaction. Diagrams A and B have the same wavelength. Diagram C has a longer wavelength than A or B.

Imagine the waves shown in the diagrams moving across the page. You can imagine that, as the wave in Diagram A or B moves, more waves will pass any one point on the page than when the wave in Diagram C moves. The number of waves that pass a point each second is called the wave **frequency**. Sound waves vibrate very fast. The faster something vibrates, the more waves it produces each second. Wavelength and frequency are related to each other. Waves with a shorter wavelength have a higher frequency than waves with a longer wavelength. Something that vibrates fast produces sound waves of a higher frequency than something that vibrates more slowly. **Pitch** is determined by the frequency of the sound wave. Sounds with a high frequency have a high pitch. Sounds that have a low frequency have a low pitch. You hear sound waves with higher frequencies, like those from a whistle, as higher pitched sounds. Other sounds, like sound waves from a bass drum, you hear as lower pitched sounds.

The diagrams are models that help you imagine sound waves, but sound waves cannot be seen. Also, most sound waves in air have a longer wavelength than those shown. The waves in the diagrams have relatively short wavelengths, which are close to the wavelengths of the highest pitch sounds humans can hear.

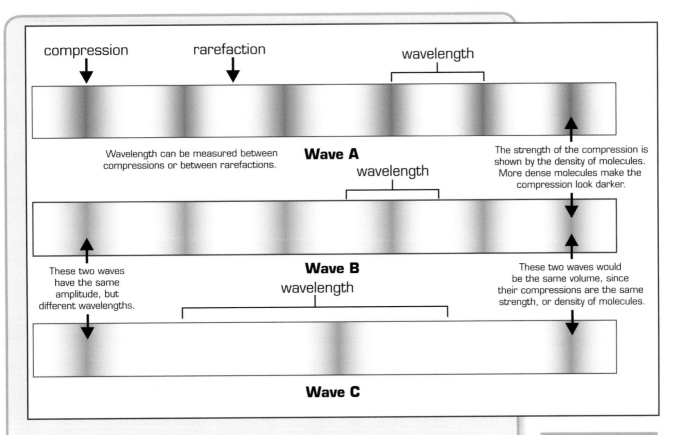

compression   rarefaction   wavelength

Wavelength can be measured between compressions or between rarefactions.

Wave A

wavelength

The strength of the compression is shown by the density of molecules. More dense molecules make the compression look darker.

wavelength

Wave B

These two waves have the same amplitude, but different wavelengths.

These two waves would be the same volume, since their compressions are the same strength, or density of molecules.

wavelength

Wave C

Frequency is measured in a unit called a **hertz**. The unit symbol for one hertz is 1 **Hz**. One hertz (1 Hz) is one wave passing a point every second. Human ears can hear sounds within a range of frequencies. Humans can typically hear high sounds with a frequency of about 20,000 Hz. This means that 20,000 waves pass a point in one second. The lowest sounds humans can typically hear are at about 20 Hz. Other animals can hear sounds humans cannot hear. The chart on page 129 shows frequencies that can be heard by different animals.

Another important characteristic of a wave is how packed together the molecules are. You can see that the molecules in A are more packed together than in either B or C. In the diagram, this is shown by the darker color, which is caused by there being more molecules packed into the space. The characteristic of how packed the molecules are is called the **amplitude** of the sound wave. Wave A has greater amplitude than either B or C. Waves B and C have the same amplitude, which can be seen in the diagrams by the compressions being the same darkness. Amplitude tells you how loud or soft the sound is. A loud sound will have greater amplitude and a soft sound will have smaller amplitude. Loudness is determined by the amount of energy used to produce the sound.

hertz (Hz): a unit used to measure frequency. For wavelength it is equal to one wave per second.

amplitude: the property of a sound wave related to how packed the molecules are.

loudness (or intensity): how loud or soft a sound is.

decibel: unit used to measure loudness.

You hear sound waves that have greater amplitude (stronger compressions) as louder sounds.

It is easy to confuse pitch and **loudness**. The two characteristics are different. Pitch is determined by frequency. Loudness is determined by amplitude. It is possible to have a sound you hear as a low pitch, like a bass guitar string, that is either loud or soft. It is also possible to have a sound you hear as high pitched, like a note from a flute, be either loud or soft. The way the string or air vibrates determines the frequency and pitch. How hard the string is strummed or the amount of air that moves through the flute determines the amplitude and loudness.

Loudness is measured in a unit called a **decibel** (db). The sound level in your classroom is probably about 35 decibels. The sound a jet engine makes is about 170 decibels at its loudest. Standing close to a jet engine may be very painful to human ears. Loud sounds can harm your ears. People who work in really loud places wear ear plugs or ear muffs to protect their ears. Otherwise, they would lose their hearing.

Sound Levels in Decibels

	0 dB	Weakest sound heard
	60 –70 dB	Normal converstaion (3–5)
	80 dB	Telephone dial tone
	85 dB	City traffic (inside car)
	90 dB	Train whistle at 500
Level at which sustained exposure may result in damage to the ears, which can decrease the ability to hear	**95 dB**	Subway train at 200
	90 – 95 dB	Power mower
	107 dB	Power saw
	110 dB	Pneumatic riveter at 4
Ears can begin to hurt from sound being too loud	**125 dB**	Jet engine at 100
	140 dB	
Ears can be so badly damaged that the tissue dies	**180 dB**	
	194 dB	Loudest sound possible within the range of human hearing

How Humans Hear

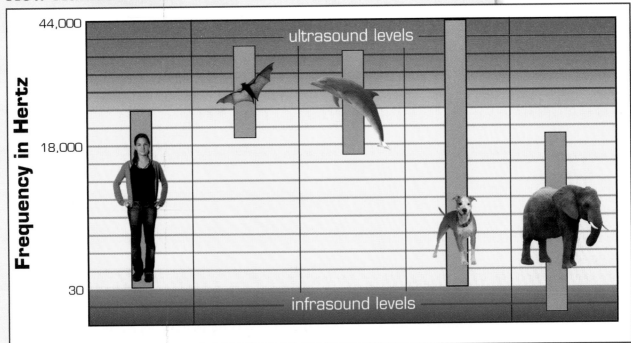

infrasound levels

Frequency in Hertz

44,000

ultrasound levels

18,000

30

Ears are the sense organs that receive sound. In the ear, vibrations are transformed into nerve impulses. Think about the outside of your ears. This is the part of your ears you are most familiar with. The outer ear is shaped like a funnel, wide at one end and narrow at the other end. The wide opening of the human ear channels sound waves into the slender tube leading into your head. In the ear, this slender tube is called the **ear canal**. The ear canal links the outer ear to the **eardrum**. The eardrum is a tightly stretched membrane similar to the surface of a drum. When sound waves reach the eardrum, the **membrane** of the eardrum begins to vibrate. This vibration is then transmitted to the middle ear.

String instruments, like violins and harps, make sounds when the strings vibrate. Thicker strings vibrate slowly and make sounds you hear as a lower pitch than the sound made by thinner strings. At the shallow end of the harp, the strings are thin and can vibrate very quickly. Those sounds you hear as higher in pitch.

ear: the sense organ involved in receiving sound waves.

ear canal: a tube-like structure connecting the external ear to the eardrum.

eardrum: the membrane separating the outer ear from the inner ear.

membrane: a layer of tissue that serves as a covering, connection, or lining.

PBIS

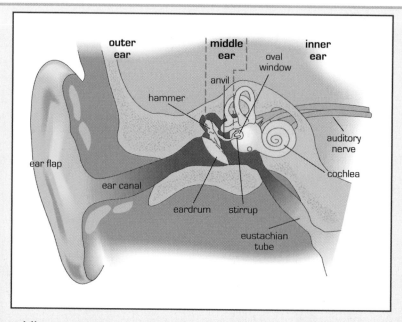

hammer: the first in a series of three small bones in the middle ear.

anvil: the second in a series of three small bones in the middle ear.

stirrup: the third in a series of three small bones in the middle ear.

oval window: an opening between the middle ear and the inner ear.

cochlea: a spiral tube that makes up the part of the inner ear responsible for hearing.

cilia (singular, cilia): tiny hairs that line the cochlea and help turn vibrations into nerve impulses.

The middle ear starts at the inside of the eardrum. Inside the middle ear are the three smallest bones in the human body. They are the **hammer**, **anvil**, and **stirrup**. Each is named for its shape. The eardrum first transmits vibrations to the hammer. The hammer then transmits the vibration to the anvil, which in turn, transmits the vibration to the stirrup. Vibrations from the stirrup are transmitted to another membrane covering an opening called the **oval window**. The oval window separates the middle ear from the inner ear.

The largest structure in the inner ear is called the **cochlea**. The cochlea is a fluid-filled tube that is spiral, similar to a snail's shell. The cochlea is lined with thousands of tiny hairs, called **cilia**. These are not like hairs on your head. They are much smaller. They are connected to cells attached to nerve endings that line the surface of the cochlea. The cilia are very sensitive to the vibrations inside the cochlea. When vibrations are transmitted to the cochlea, the fluid vibrates, pushing the tiny

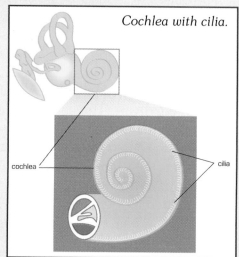

Cochlea with cilia.

cilia back and forth. The movements of the cilia are turned into nerve impulses. These nerve impulses travel along the auditory nerve to the part of the brain responsible for interpreting sound.

3.6 Explore

What Affects How Marine Mammals Communicate?

Animals communicate through sound, sight, touch, and smell. The puzzle activity you did, using verbal and non-verbal communication, showed the advantage humans have in using spoken language. However, you also observed some fascinating examples of how other animals can exchange important information using their own forms of communication.

The means of communication used by elephants and bees depends largely on their environment. Elephants can send sounds over long distances to keep in contact with other members of their extended family. Scientists think bees perform a waggle dance on the surface of their hive to communicate the distance and direction of a food source. There are many animals, however, that live in a different type of environment. They are **marine mammals,** or mammals that live in seawater. Sometimes their world is a dark and murky underwater environment. At other times, they are on the surface of the water where there is plenty of sunlight.

Marine mammals, like dolphins and whales, live in groups, called **pods**. They swim long distances through the ocean in these groups. They rise to the surface of the water to breathe oxygen, and then dive deep into the water to feed. Like elephants, bees, and people, marine mammals need to communicate to their group members their location and the location of food.

marine mammal: a mammal that lives in the sea and/or gets its food from the sea.

pod: a social group of whales or dolphins. Members of a pod may protect one another.

Dolphins live in groups, called pods.

Predict

Dolphins must be able to communicate on the surface and underwater.

Working by yourself, think about where marine mammals live and the reasons why they need to communicate. Make a list of marine mammals' communication needs. Then get together with your group and develop a group list of marine-mammal communication needs.

Work together with your group to predict how you think dolphins might communicate with other dolphins. Remember that environment plays an important role in communication. Identify the constraints the dolphins' bodies and environment place on dolphins' forms of communication. Consider the fact that sometimes dolphins swim at the top of the water, where there is plenty of sunlight, and sometimes they swim much deeper in the ocean, where it can be very dark and hard to see.

Observe Dolphin Communication

Once again, you will watch a video to observe animal behavior. You will watch a video showing dolphins communicating. As before, you will watch the video twice. The first time, you will be watching it to come up with an observation plan. The second time, you will watch more carefully, using your plan, and then work with your group to identify all the different forms of dolphin communication.

A baby dolphin relies on touch and sound to remain close to its mother.

Watch the video. Pay attention to the behavior of the dolphins. Try to figure out what they are doing and what behaviors might be communication. Notice if all the dolphins are behaving the same way.

Plan

You watched a video of dolphins communicating. When you watch the video again, you will need to make detailed observations about how dolphins communicate. Make an observation plan with your group. When you observed elephant communication, you also developed a plan. You might use a similar plan this time. Create your plan, keeping in mind that you need to observe as much of the scene as possible, and that your observations will determine how well you understand the communication of dolphins. Make sure that part of your plan includes using a page to record your notes.

Observe

Once you have your plan written and all the members of your group understand the plan, watch the video again. Remember to take notes about the dolphins' behavior and habitat. Pay attention to the dolphins' bodies. Think about how their body structure helps them to communicate. How does a dolphin's habitat affect how it communicates? Remember to follow the plan your group decided on. Record your observations so you will be able to share them with your group and your class.

Analyze Your Data

The video you watched showed a group of dolphins. How did you observe them communicating? You may have predicted and observed that dolphins make different sounds. These sounds communicate a variety of things to other dolphins. This is one way other mammals, such as people, communicate. You may also have seen the dolphins communicate in other ways.

Meet with your group to share your observations. Each member of the group should have a chance to share all their observations. Remember that, depending on your observation plan, you may not have been watching what the other group members were watching. Other group members may have a lot of information to share with you. Listen carefully for observations you might not have seen. Create a group list of observations that *everyone* agrees on. Make another list of observations that were not agreed on. Save the observations you could not agree on for the *Investigation Expo*.

Using the same procedure you used while analyzing the behavior of elephants (in *Section 3.4*), create a sticky note for each observation you all agreed on. Organize your observations into groups based on the type of communication behavior you observed. For example, you might have seen dolphins touch. If you think that some behaviors might fall into two groups, you can list those behaviors twice.

It is time to think about what your observations mean. Think about when dolphins make different sounds, and what the different sounds might mean. Think about the different ways dolphins use their bodies to send information to each other, and what those body movements might mean. These interpretations will help you as you begin to develop explanations. Using an *Observing and Interpreting Animal Behavior* page, record the observations your group agreed on in the left column. Include one behavior in each row of the chart. Then add to the middle column what you know about what allows these behaviors, and record your interpretations of these behaviors in the right column. Collaborate with the members of your group to create one chart. Be prepared to share your observations and interpretations with the class.

Communicate

Investigation Expo

You will share your observations and data analysis in an *Investigation Expo*. In your presentation, you will share what you have seen with the class. Remember that each group created their own plan for making their observations and analyzing their data. Some groups' observations and interpretations may be different from yours.

For your presentation, create a poster that describes each of the following:

- the questions you were trying to answer in your observations
- your observation procedure and how it helped you make observations
- your analysis of your observations and how confident you are about the analysis

Add to the bottom of your poster the list of observations your group did not agree on.

Before the presentations, read these questions. Plan your group's presentation, making sure you will answer all of the questions. Then as each group is presenting, listen for the answers to these questions. If a group does not answer all the questions, ask them to help you understand what they observed and why they interpreted behaviors the way they did. When you ask your questions, focus on better understanding the observations the group made.

- What was the group trying to find out?
- What procedure did they use to collect their data?
- Were they able to make clear observations that were detailed?

- How did they group their observations? What did their groupings allow you to see?
- What conclusions do their results suggest?
- Do you trust their observations? Why or why not?

Explain

Once again, using your observations and interpretations, create an initial explanation of what affects dolphin communication. Your explanation of dolphin communication behavior must bring together the evidence (observations), the interpretations the class agreed on, and your science knowledge. Using a *Create Your Explanation* page, develop your explanation of why dolphins communicate as they do. Remember that the best explanations help others understand what makes a claim valid. At this point, your science knowledge may be limited. In the next section, you will read more about dolphins, and you will have an opportunity to improve your explanations.

Communicate

Share Your Explanation

Share your explanation with the class. As each group shares their explanation, pay special attention to how the other groups have supported their claims with science knowledge. Ask questions or make suggestions if you think a group's claim is not as accurate as it could be or if the group has not supported their claim well enough with observations and science knowledge. Remember that the best explanations help others understand what is true about the world that makes the claim trustworthy.

What's the Point?

Marine mammals, such as dolphins, communicate in very different ways than other animals. Dolphins live in a different type of environment than humans, bees, and elephants. The ocean can be a dark and murky place deep underwater or a sunny place near the surface. Dolphin communication must adapt to the type of environment.

While observing dolphins communicating, once again it is important to make observations first then interpret the observations. This is how scientists study animal behavior. They are aware of the difference between observations and interpretations.

3.7 Read

How Do Dolphins Communicate?

Sometimes, when people cannot see each other but know that someone is close enough to hear, they use their voices. The sound of a human, or an elephant, travels well through the air. When marine mammals communicate, the sounds they make must be able to travel well through the water. Marine mammals have developed many different ways to communicate in a water environment.

How Do Dolphins Use Sound to Communicate?

To survive, dolphins must be able to keep in touch with other members of their pod, identify and avoid obstacles and predators in the ocean, and find food. Their environment requires that they communicate in ways other than just visual communication. The ocean can be dark and murky, and finding food, ocean hazards, or other dolphins through sight alone is not always possible.

Dolphins must be able to communicate in dark, murky areas of the ocean.

Dolphins use clicks, whistles, squeaks, and trills to communicate. These kinds of sounds travel well through water. One specific type of click is called a "*sonar* click." (Sonar stands for "SOund NAvigation and Ranging." You will read more about sonar later in this section.) Sonar clicks allow dolphins to communicate with **echolocation**. Echolocation works just like an echo. When a dolphin makes a sound in the water, the sound waves move through the water and hit an object. The sound waves then bounce off the object and travel back to the dolphin. The dolphin hears the returning sound wave. The time it takes for the sound wave to travel out and come back gives the dolphin information about the size, shape, speed, distance, and direction of objects in the water. Echolocation is very accurate. Dolphins are so good at using echolocation that they can even use it to tell the difference between types of fish.

In water, the molecules are closer than they are in air, therefore sound travels almost five times faster in water than it does in air. This helps dolphins receive information more quickly than if they were sending the sounds through the air.

echolocation: a method used by dolphins and some other animals such as bats, to locate objects. The animal sends out sound waves that bounce off the object. The returning sound (echo) is interpreted to determine the shape and the location of the object.

Marine mammals have excellent hearing. The ocean is a noisy place. Many creatures make sounds. People have also added many sounds to the ocean. Dolphins can distinguish those sounds from one another. They can hear the echoes of their own clicks, and they can hear the clicks of other dolphins. They can use the clicks of other dolphins to find them. By using their excellent hearing, dolphins can find food, avoid obstacles, find other dolphins, and avoid some dangers.

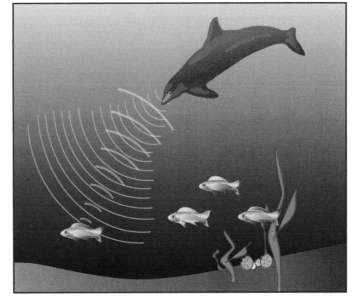

Dolphins use echolocation to find fish.

Dolphins use sound in other ways, too. Each individual dolphin has its own "signature whistle." A mother dolphin will repeat this signature whistle over and over to her newborn. The calf becomes able to identify its mother's special sound. When a dolphin mother is separated from her calf, she whistles her sound. Because the calf can recognize the sound, it knows where to find its mother.

What Are Some Other Ways Dolphins Communicate?

Dolphins are mammals and must breathe air to survive. To get air, they must return to the water surface. In the brighter ocean near the surface, dolphins often use sight and touch, as well as sound, to communicate with one another. Dolphins can be seen leaping high into the air. They also slap their flippers or tails on the surface of the water. Using this body language, dolphins can alert others to danger and possible food sources, or can tell others they want to play.

*Dolphins often jump high above the water surface and fall back with a loud splash. This is called **breaching**. Scientists interpret this behavior as a form of communication.*

Touch is an important part of communication for dolphins. Scientists have observed dolphins rubbing, petting, or hitting one another. Often, they maintain contact with one another while they are swimming. When dolphins meet, they may rub fins.

breaching: a behavior seen in marine mammals and some fish, where the animal jumps high above the water surface and falls back with a splash.

sonar: a technique that uses sound to provide images of objects that are under water.

oceanographer: a scientist who studies the ocean.

Stop and Think

1. What are some ways dolphins communicate with one another?

2. People communicate about a variety of things. What do you think are some things dolphins communicate about?

3. How does a dolphin's environment affect how it communicates?

Humans also Use Echolocation

Echolocation is a very important tool for marine mammals and some other mammals, such as bats. Scientists also use echolocation, called **sonar**. **Oceanographers** are scientists who study the ocean.

Project-Based Inquiry Science

Because parts of the ocean are very deep and dark, these scientists cannot explore it themselves. However, they can make accurate images of what the ocean floor looks like using sonar. This allows them to study underwater mountains and ridges. Sonar use also allows oceanographers to find sunken ships.

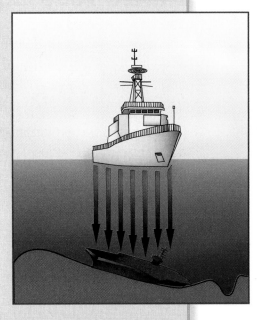

Scientists use sonar to map the floor of the ocean.

As mentioned, sonar stands for "SOund NAvigation and Ranging." Scientists send out sound waves from the water and measure the time it takes for each one to return. Each measurement is recorded. When the measurements are put together, they create an image of what is on the ocean floor. Sonar has expanded scientists' ability to see where they have never seen before.

Sonar can also be used by ships to see other ships, find obstacles in the water, and even find fish. Since World War I, in 1915, nations have used sonar on naval ships to find and intercept enemy vessels. Unfortunately, these uses of sonar have been linked to deaths of marine mammals around the world.

Although scientists do not understand why, it is clear that some types of sonar used by these ships interfere with the echolocation of marine mammals. Scientists have found several cases of whales having beached themselves after having been exposed to the sonar of ships.

*One of 12 sperm whales that **beached** and died on Karekare Beach, West Auckland, New Zealand, in late 2003.*

beached (beaching, to beach): when a marine mammal that cannot live out of the water strands itself on land, usually a beach.

ANIMALS IN ACTION

Revise Your Explanation

You have just read more about how dolphins communicate and what scientists know about dolphin communication. With your group, look back at your *Observing and Interpreting Animal Behavior* pages. Look at your original interpretations of dolphin communication. Now, with your new science knowledge, reinterpret the behavior you saw in the video.

Go back to your explanation on your *Create Your Explanation* page. First, add the science you just learned to the science knowledge box. Then, check to make sure your claim is still accurate. If your claim does not match the science you have read, revise it. Next, support your claim with your new science knowledge.

Rewrite your explanation to make it more complete. Include in your explanation how the dolphin's body structure and environment influence its communication. The dolphin's environment includes the dark area under water as well as the sunlit area on the surface of the water. Be sure to include both of these areas in your explanation. Remember that an explanation is a statement that connects a claim to evidence and science knowledge in a logical way. Write your explanation so that it tells why your claim is accurate. Be sure your explanation matches the science you just read. Make sure your claim now matches what you have learned. If it does not, revise your explanation. Use the information from your reading about dolphin communication to support your revised explanation. You might need to write an explanation that has a few sentences rather than just one long sentence. The goal is to tie everything together and help others understand why your claim is true.

Communicate

Share Your Explanation

Share your new explanation with the class. When you share your explanation, tell the class what makes this revised explanation more accurate than your earlier one. As each group shares their explanation, pay special attention to how the other groups have supported their claims with science knowledge. Ask questions or make suggestions if you think a group's claim is not as accurate as it could be or if the group has not supported their claim well enough with observations and science knowledge.

As a class, work together to develop your best explanation of what affects how dolphins communicate.

Update the *Project Board*

As you read more about how dolphins communicate, you may have thought of new questions you want to ask. Add what you learned from your reading to the *Project Board*. Make sure you add evidence to support any new science you learned. Also, think about how this learning will help you answer the *Big Question, How do scientists answer big questions and solve big problems?* As your teacher records this information on the big *Project Board*, add the information to your own *Project Board* page.

What's the Point?

Echolocation and sonar both rely on sound waves being sent out, bounced off objects, and returned. The way the sound comes back helps determine the shape and location of an object. Dolphins use echolocation. Sets of sonar clicks make it possible for a dolphin to find food, other dolphins, and hazards in the water. Echolocation is very important for dolphins because ocean water can be murky and dark. Other forms of communication are not as effective in that environment.

Sonar is also a tool used by scientists to "see" into the deep and dark parts of the ocean. Sonar uses sound waves to create an image of the ocean floor, fish, and objects in the ocean, such as sunken ships. Naval vessels use sonar to find and intercept other ships. Scientists think there are times when the sonar from these ships interferes with the echolocation of marine mammals.

This dolphin is using echolocation to identify an object inside a box.

Learning Set 3

Back to the Big Challenge

You have been developing an answer to the question, *What affects how animals communicate?* Answering this question will help you complete the challenge of the Unit, to design an enclosure for an animal. Your enclosure needs to be designed so that your animal can communicate as naturally as possible. The enclosure needs to be as similar as possible to the animal's habitat.

To develop your recommendations about the design of an animal enclosure, you will need to think about some of the big ideas that affect how animals communicate. Several big ideas were introduced in this *Learning Set*, and you need to think about how those ideas are going to be built into your enclosure.

You are going to focus on one of the animals you learned about in this section. You need to use the explanations regarding that animal to help you develop recommendations about designing your enclosure. You can also use information from the *Project Board*, especially the third and fourth columns where you have included information about each animal. You will then share your recommendations with the class.

Develop Recommendations

A recommendation is a type of claim. You will support each of your recommendations with evidence from your observations and reading. Some of this evidence may come from your previous explanations. Some will come from your observations and interpretations. To prepare for presenting to the class, and so that you can be sure that your recommendations match the evidence you have collected and what science tells you, use a *Create Your Explanation* page for each recommendation your group makes. Your recommendation will be your claim. Add evidence and science knowledge that supports it. Then develop a logical statement linking your recommendation to the evidence and science knowledge.

You will write a set of recommendations for how to provide the best environment for some animal so it can communicate in the most natural way possible. Think about starting your recommendations with "If," "When," or "Because." For example, you might begin your recommendation by writing, "If the dolphins were going to communicate in the most natural way possible. . . Even better would be a recommendation of the form, "Because dolphins need to. . . " Then you need to complete the statement.

It will be necessary to create more than one recommendation. You have learned a lot in this *Learning Set*, and you want to show what you have learned. You will need to write as many recommendations as you can about how an enclosure will ensure that your animal is able to communicate naturally.

Communicate Your Solution

Solution Briefing

After you have developed your recommendations, you will communicate your recommendations to one another in a *Solution Briefing*. In a *Solution Briefing*, you present your solution, or recommendations, in a way that will allow others to evaluate how well it achieves the criteria and to make suggestions about how you might improve it.

As you prepare for this briefing, make sure you revisit the criteria and constraints you identified in the beginning of the Unit. Use the following questions to plan your presentation.

- How is your enclosure addressing the communication needs of your selected animal?

- How does the enclosure meet the criteria?

- How did the constraints affect your recommendations?

- What information did you use to help you make each recommendation?

- What other ideas did you think about along the way, and why did you not recommend them?

- What questions do you still have?

As you listen to the presentations, make sure you understand the answers to these questions. If you do not understand something, or if something is not presented clearly enough, ask questions.

Solution-Briefing Notes				

Name:_____ Date:_____

Design Iteration: _____

Design or group	How well it works	What I learned and useful ideas		
		Design ideas	Construction ideas	Science ideas

Plans for our next iteration

You can use the questions above as a guide. When you think something can be improved, make sure to contribute your ideas. Be careful to ask your questions and make your suggestions respectfully. As you listen, record notes on a *Solution-Briefing Notes* page.

Update the *Project Board*

The last column on the *Project Board* helps you pull together everything you have learned during the Unit. You can then use what you record there to address the *Big Challenge* and answer the *Big Question*. Each investigation you do and reading you complete is like a piece of a puzzle. You fit the pieces together to help you address the challenge. Each piece provides you with a critical factor that must be addressed to answer the *Big Question*.

The *Big Challenge* is to create an animal enclosure that makes it possible for the animal to feed or communicate as naturally as possible. The last column of the *Project Board* is the place to record your knowledge that might help you address this challenge. This column is also the place to record recommendations about how to address the challenge.

Now, add your recommendations to the *Project Board*. By adding this information, you will be able to save it for when you return to the *Big Challenge* at the end of the Unit.

Address the Big Challenge

Design an Enclosure for a Zoo Animal that Will Allow it to Feed or Communicate as in the Wild

Your challenge for this Unit is to design a zoo enclosure that will accommodate the feeding or communication of one of the animals you studied in the Unit. Your goal will be to design the zoo environment so it is similar enough to the natural environment of the animal to allow the animal to feed or communicate effectively. The enclosure will also have to allow visitors and scientists to observe the animals clearly.

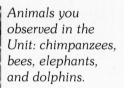

Animals you observed in the Unit: chimpanzees, bees, elephants, and dolphins.

You now know enough to address the challenge. You observed the behavior of several animals in the Unit. You analyzed their behavior, identified why they were behaving the way they were, and recorded your data on *Observing and Interpreting Animal Behavior* pages. These pages will be useful in designing your enclosure. They contain the behavior you saw and how you interpreted that behavior. The details you entered may help you identify how your animal's environment affects its behavior. You also developed explanations of why the different animals behave the way they do. As a class, you developed recommendations about designing enclosures for each of the animals. You recorded your claims and recommendations on *Create Your Explanation* pages and included evidence and science knowledge to support your claims. All of this will be useful to you as you address the challenge.

Identify Criteria and Constraints

The first thing designers do when they are asked to address a challenge is to identify the criteria and constraints. You began the Unit by identifying some of the criteria and constraints for this challenge. But you now know a lot more about animal behavior than you did at the beginning of the Unit. You also now know which animal you will be planning an enclosure for and whether you will be focusing on the animal's feeding or communication.

As you develop a more complete set of criteria and constraints for your design, keep the questions in the table in mind.

Questions for those addressing feeding needs	Questions for those addressing communication needs
How large is your animal?	How large is your animal?
What size group does it live with?	What size group does it live with?
What kind of climate does it live in?	What kind of climate does it live in?
What kinds of plants and animals live in its habitat?	What kinds of plants and animals live in its habitat?
Are there other special features of its habitat that are important to its feeding behavior?	Are there other special features of its habitat that are important to its communication behavior?
What foods does it eat?	For what purposes does it communicate?
Where does it find its food?	How does it send messages?
How does it obtain its food?	How does it receive messages?

As you answer the questions in the table and identify criteria and constraints, listen to each other's ideas. It is important for everyone's answers and ideas to be heard.

Each animal has different needs. You must keep those needs in mind as you plan the enclosure for your animal. Your next step is to identify the criteria and constraints you will need to keep in mind as you design an enclosure for your animal. For example, if your animal eats leaves from the tops of tall trees, one of your criteria will be to include those kinds of trees in the enclosure. If your animal lives in large groups, one of your criteria will be that the enclosure has to be large enough for a natural group.

Plan Your Design

You have learned about the ways each animal's body and habitat affect its ways of finding food and communicating. As you design your enclosure, keep these constraints in mind. Remember to use the *Project Board* and your interpretations, explanations, and recommendations as resources as you design your enclosure.

Make decisions about the design of the enclosure together with your group. For example, how much space will you need and why? How will you make sure your animal will be able to use its body the way it does in the wild? What plants and trees will the habitat need, and why? What other animals will the habitat need to include, and why? How will your animal be able to get the food it needs? How will you make sure it can communicate the way it needs to? For each decision you make, discuss alternatives with your group. Know why you are making each decision, and record the evidence and science knowledge you are using.

When you are finished, you will have a chance to share your plan with other groups in a *Plan Briefing*. Others in the class might be able to help you with any difficult decisions you need to make as you work on your plan. You will get a chance to iterate on your plan and improve it based on others' feedback.

Communicate Your Design Plan

Plan Briefing

As you are finishing your plan, begin to draw a poster for presentation of your plan to the class. Include in that plan a drawing of your enclosure. Be sure to label the details of the enclosure. Be prepared to describe all the parts of your plan to others and to support each of the parts of your plan with evidence and science knowledge.

Introducing a *Plan Briefing*
Preparing a *Plan-Briefing* Poster

A *Plan Briefing* is much like the other briefings you have engaged in. In a *Plan Briefing*, you present the plan you are developing. You must present it well enough so that your classmates can understand your ideas and why you made each of your decisions. They should be able to identify if you have made any mistakes in your reasoning. Then they can provide you with advice that you can use to improve your plan. As a presenter, you will learn the most from a *Plan Briefing* if you can be very specific about your plan and about why you made your plan decisions. You will probably want to draw pictures, maybe providing several views. You certainly want everyone to know why you expect your plan to achieve the challenge.

The following guidelines will help you as you decide what to present on your poster.

- Your poster should have a detailed sketch of your plan with at least one view. You might consider sketching multiple views so that the audience can see your plan from different angles. It is important that the audience can picture your design.

- Parts of the plan and any special features should all be labeled. The labels should describe how and why you made each of your plan decisions. Show the explanations and recommendations that support your decisions. Convincing others that your plan choices are quality ones means convincing them that you are making informed decisions backed by scientific evidence.

- Make sure to give credit to groups or students who gave you information or ideas for your plan. If another group provided an Explanation or a Recommendation you are using, you should credit them with their assistance in developing your final plan.

Participating in a *Plan Briefing*

As in other presentation activities, groups will take turns making presentations. After each presentation, the presenting group will take comments and answer questions from the class.

When presenting, be very specific about your plan and what evidence helped you make your plan decisions.

Your presentation should answer the following questions:

- What are the critical features of the plan?
- For each feature, what criterion will it achieve? Why is this way the right way to achieve that criterion?
- What issues are you still thinking about?
- How did you use the explanations and recommendations the class developed to help you with your plan?
- In what ways do you need the help of other groups? What issues can they help you solve?

As a listener, you will provide the best help if you ask probing questions about the things you don't understand. Be respectful when you point out errors in the reasoning of others. These kinds of conversations will also allow listeners to learn.

For each presentation, if you don't think you understand the answers to the questions above, be sure to ask questions. When you ask others to clarify what they are telling you, you can learn more. They can learn, too, by trying to be more precise.

Update Criteria and Constraints
Revise Your Plan

You will now have some time to work further on your plan based on suggestions your classmates made in the recent briefing. As you listened to the presentations of others, you may also have thought about other things you want to add to your plan. Each change and new plan is called an iteration.

You may have received some good advice during the *Plan Briefing*. Others may have made suggestions about ways to make your plan better. You may also have discovered new criteria and constraints that you were not aware of earlier. You will have a chance to update your criteria and constraints and to revise your plan. As you do that, be sure to update the pages you are using to record your plan and to justify your decisions. Be sure to revise your sketches and the labels on them.

As you think about which of your classmates' suggestions to include in your new plan, think about whether each suggestion is valid, given what you know about your animal.

As you update your criteria and constraints and then your plan, keep in mind one further issue. Observers need to be able to make good observations of your animal in the enclosure. Consider the type of enclosure and the type of observations the observers might want to make. To create an enclosure that will allow good viewing, you will need to answer the following questions.

- Where in your enclosure would the observer be able to observe feeding and communication?

- What type of data would an observer like to collect?

- Where would an observer sit or stand to get the best view?

- Would all the observer's observations be made with her or his eyes?

- How would the observer use her or his ears to make observations?

- What special instruments might the observer use to gather data?

- Where might those instruments be located in the enclosure?

Communicate Your Plan

Plan Briefing

In this *Plan Briefing*, you will focus on the revisions you made to your plan to allow visitors to observe your animals. You may present a revised poster, or you may need to make a new one.

When presenting, be very specific about your revised enclosure plan and what evidence helped you make your new decisions. Also, make sure you give credit to the groups that helped you think about how you might plan your enclosure. If another group provided an explanation or evidence you are using, credit them with their assistance in developing your plan.

Your presentation should answer the following questions.

- What are the critical features of the revised plan?

- What criterion of the challenge will each feature achieve? What makes the revisions you made to your plan a better way to achieve that criterion?

- What issues are you still thinking about?

- In what ways do you still need the help of others groups? What issues can they help you solve?

For each presentation, if you do not think you understand the answers to the above questions, be sure to question your classmates. When you ask them to clarify what they are telling you, you can learn more. They can learn, too, by trying to be more precise.

Revise Your Plan

With your group, take into account the advice of your classmates, and revise your plan one last time.

Be a Scientist

Copying versus Crediting

When you build on someone else's idea, it is important to give them credit. Why isn't this "copying"? Copying means taking the work of someone else and claiming it as your own. If you simply sketch what some other group sketched, that is copying. But if you add to another group's idea and acknowledge from where you got your idea, you are doing what scientists do. When you explain how you used their ideas and made them better, you are adding your contribution to theirs.

This means you need to keep good records of where you get your ideas. When you use someone else's ideas, always record from whom you borrowed the idea. Record how you included it in your design and why you did it that way. Then make sure to give credit to the other person or group in your presentations.

Communicate Your Solution

Solution Showcase

After every group has a chance to iterate on their plans, it will be time to complete this challenge. You will present your final design in a *Solution Showcase*.

The goal of a *Solution Showcase* is to have everyone better understand how each group approached the challenge. You will see several solutions for the challenge, each designed to accommodate a specific animal's needs.

Be sure to discuss how you included in your final design the explanations and recommendations that the class generated. Explain why you think your solution is a good one.

Your presentation should show a picture of your animal enclosure. The picture should include as much detail as possible, and all the critical parts should be labeled. Be prepared to carefully describe how the features of your enclosure will help meet the needs of your animal for feeding or communication. Be sure to give credit to others who helped you improve your plan.

Your presentation of your enclosure design should show

- how your animal will feed or communicate naturally

- how you accounted for criteria and constraints of the challenge

- how you took into account any special needs your animal might have

- how observers will view the behavior of your animal

Be a Scientist

Introducing a Solution Showcase

The goal of a *Solution Showcase* is to present a completed solution and help everyone understand how each group arrived at their solution. You have the opportunity to see how each group took into account the advice given earlier.

Your presentation during a *Solution Showcase* should include the history of your plan. Review your original plan. Then tell the class why and how you revised it. Make sure to present the reasons you made the changes you did. Do this for the whole set of iterations you did. Make sure too, that the class understands what your final plan is. As you prepare, you will need to organize your thoughts so you can present the history of your plan quickly and completely.

As you listen, it will be important to look at each plan carefully. You should ask questions about how each plan meets the criteria of the challenge. Be prepared to ask (and answer) questions such as these.

- What approaches were tried and how were they done?
- How well does the solution meet the goals of the challenge?
- How did the challenge constraints affect your solution decisions?
- What problems remain?

Update the *Project Board*

Now that you have completed your challenge, you will go back to the *Project Board* for one final edit. You will focus mainly on the middle column, *What are we learning?* and the last column, *What does it mean for the challenge or question?* Record what you have learned about what animals need to feed and communicate, and how this applies to designing a natural enclosure.

What's the Point?

To help in designing zoo enclosures, ethologists study animals and how animals behave. Addressing this challenge requires a lot of thinking. The scientists who design zoo enclosures must think about the behavioral and environmental needs of the animal. In many cases, they must also think about how other scientists can observe the animal in its enclosure. They aren't always able to fully design a proper enclosure the first time.

By observing animals, listening to the advice and ideas of other scientists, and using that information to improve upon their original design, scientists can build the best enclosure for an animal.

Answer the Big Question

How Do Scientists Answer Big Questions and Solve Big Problems?

You addressed the *Big Challenge* by planning, sketching, and sharing with others an enclosure for an animal. The Unit also includes a *Big Question*. The *Big Question* is *How do scientists answer big questions and solve big problems?* Like scientists, you planned, recorded, analyzed, and explained your animal observations. You also read about several ethologists who study animal behavior. The following questions will help you remember what you now know about how scientists answer big questions. Your answers to these questions will help you answer the *Big Question*. Write an answer to each question. Use examples from class to justify your answers. Be prepared to discuss your answers in class.

1. **Teamwork**—Scientists often work in teams. Think about your teamwork during the Unit. Record the ways you helped your team. What were you able to do better as a team than you could have done alone? What things made working together difficult? What did you learn about working as a team?

2. **Learning from other groups**—What did you learn from other groups? What things did you help other groups learn? What is required to learn from another group or help another group learn? How can you make *Plan* and *Solution Briefings* work better?

3. **Iteration**—You used iteration to improve your explanations and your solutions to the challenge. Iteration is more than simply trying again. What is iteration helpful for? How did iteration help you with observations and explanations? How did it help you design your enclosure?

4. **Meeting criteria and dealing with constraints**—What are criteria? What are constraints? How is specifying criteria and constraints useful for solving problems and addressing design challenges?

5. **Using cases to learn**—It is interesting to learn how scientists do their work. In the Unit, you learned about Jane Goodall, Karl von Frisch, and Alfred Wenner. They each showed you a little about how scientists work. You worked as an ethologist in the Unit in ways similar to these ethologists. What are the benefits of learning how other scientists work?

6. **Interpretation**—Scientists observe and then interpret. What is the difference between observation and interpretation? Why is it important to observe before you interpret? Why is it important not to confuse observation and interpretation?

7. **Explaining**—All scientists attempt to explain the way the world works. Ethologists explain animal behavior. What are the parts of a valid explanation? What makes some explanations better than others?

8. **Supporting decisions with evidence**—Scientists seek evidence so they can explain the way the world works. Ethologists' evidence often comes from their observations of animal behavior. Why is evidence so important for making good recommendations?

Glossary

adaptation
a special trait that allows an animal to survive in its environment.

amplitude
the property of a sound wave related to how packed the molecules are.

androecium
see stamen.

angiosperm
one of the major groups of modern flowering plants. They produce seeds in specialized reproductive organs called flowers.

anthers
the parts on the stamen of a flower where pollen is produced.

anvil
the second in a series of three small bones in the middle ear.

behavior
an animal's response to its environment.

biologist
a scientist who studies living things.

**breached
(breaching, to breach)**
when a marine mammal that cannot live out of water strands itself on land, usually a beach.

breaching
a behavior seen in marine mammals and some fish, where the animal jumps high above the water surface and falls back with a splash.

calf (plural, calves)
the young of certain kinds of animals, such as an elephant or a cow.

calorie
the amount of energy needed to raise the temperature of one gram of water by 1°C.

Calorie
the amount of energy in foods. One Calorie is the same as 1 kilocalorie or 1000 calories.

carnivore
an animal that eats only meat.

carpel (or pistil)
the female reproductive organ of a flower. It may be made up of a single carpel, or of two or more carpels united.

category
a set or class of things with similar characteristics, properties, or attributes.

cilia (singular, cilium)
tiny hairs that line the cochlea and help turn vibrations into nerve impulses.

cochlea
a spiral tube that makes up the part of the inner ear responsible for hearing.

collaborate
to work together.

colony (plural, colonies)
a group of similar organisms living or growing together.

compression
part of a sound wave where molecules are pushed together.

cone
a type of cell found in the retina. Each cone is sensitive to one color: red, green, or blue.

constraints
factors that limit how you can achieve a challenge.

corolla
the collection of the colored petals of a flower.

criteria (singular criterion)
goals that must be satisfied to successfully achieve a challenge.

data (singular datum)
recorded measurements or observations.

decibel
unit used to measure loudness.

ear
the sense organ involved in receiving and processing sound waves into nerve impulses.

ear canal
a tube-like structure connecting the external ear to the eardrum.

eardrum
the membrane separating the outer ear from the inner ear.

effort force
the force applied to a lever.

electromagnetic radiation
a wave that travels through space and carries energy.

electromagnetic spectrum
the range of wavelengths emitted by the Sun.

enzymes
organic substances that cause chemical changes in other substances.

ethnographer
a scientist who studies people.

ethogram
a table used to record observations of animal behavior.

ethologist
a biologist who studies the behavior of animals in their natural environment.

explanation
a statement that connects a claim to evidence and science knowledge.

fertilize
in biology, to unite male and female sex cells to form a new organism.

forage
to search for something, especially food and supplies.

forager
an organism that searches for food.

frequency (of waves)
the number of waves that pass a point in a second.

hammer
the first in a series of three small bones in the middle ear.

hatchling
a very young baby bird.

herbivore
an organism that eats only plants.

hertz (Hz)
a unit used to measure frequency. For wavelength, it is equal to one wave per second.

homeostasis
the maintenance of stable internal conditions in an organism.

instinctive behavior
a behavior an animal is born with.

interpretation
an explanation or establishment of the meaning or significance of something.

iteration
a repetition that attempts to improve on a process or product.

joint
the place where two bones come together.

larvae (singular, larva)
the newly hatched form of an insect (in this case, the bee).

learned behavior
a behavior that comes from teaching or experience.

lever
a simple machine made up of a rigid bar and fulcrum, or pivot point.

load
the object being moved, or that resists the motion of the lever.

loudness (or intensity)
how loud or soft a sound is.

machine
a device that makes work easier.

mammal
a warm-blooded animal with hair and in which the female has special glands to feed milk to its offspring

mammary glands
milk-producing glands found in female mammals that are used to feed the young.

marine mammal
a mammal that lives in the sea and/or gets its food from the sea.

metabolism
the combination of chemical reactions through which an organism builds up or breaks down materials converting energy to carry out its life processes.

membrane
a layer of tissue that serves as a covering, connection, or lining.

model
a representation of something in the world.

movable joint
joints that move freely.

mutualism
A relationship between organisms of two different species in which each member benefits.

nectar
a sugary liquid produced by plants.

nectary
the organ of plants that secretes nectar. The nectary is normally found inside the flower.

oceanographer
a scientist who studies the ocean.

observe
use one of the five senses to gather information about an object or phenomenon.

ommatidia
one of the structural elements that make up the compound eye of insects.

omnivore
an organism that will feed on many different kinds of food, including both plants and animals.

oval window
an opening between the middle ear and the inner ear.

ovary
in a plant, the enlarged part at the base of the carpel that contains ovules, or eggs.

ovule
a tiny, egg-like structure in flowering plants that develops into a seed after pollination.

pistil (or carpel)
the female reproductive organ of a flower. It may be made up of a single carpel, or of two or more carpels united.

pitch
a measure of the frequency of the vibration of the source of a sound.

pod
a social group of whales or dolphins. Members of a pod may protect one another.

pollen
small, powdery grains that contain the male sex cells in seed plants.

pollen tube
the slender tube formed by pollen grains to reach and fertilize the ovules.

pollination
the transfer of pollen (male sex cells) from an anther to a stigma of a flower.

pollinator
an insect or other animal that carries pollen from one flower to another.

predator
an organism that captures and eats part or all of another organism.

proboscis
a slender tubular feeding and sucking organ of certain insects, like butterflies and moths.

rarefaction
part of a sound wave where molecules are spread apart.

receptacle
the base of a flower to which the flower parts are attached.

receptor cells
the cells that receive information from the world and send it to the brain.

recommendation
a claim that suggests what to do in a described situation.

regurgitate
to bring partially digested food up from the stomach into the mouth.

reproduce
to produce offspring.

retina
a membrane sensitive to light that lines the back portion of the inside of the eye.

rod
a type of cell found in the retina. Each rod is sensitive to low levels of light, but cannot see colors.

roosts
communal resting places, mostly for a single species of birds.

savanna
a grassland area with scattered trees, usually found in tropical or subtropical regions.

sepals
green leaf-like parts of a flower found outside the petals.

simulation
use of a model to imitate, or act-out, real life situations.

sonar
a technique that uses sound to provide images of objects that are under water.

sound wave
the movement of molecules in a pattern, repeated over and over again, very rapidly.

species
all of the living things of one distinct kind with common characteristics, such as dogs.

spectrum
a band of colors, from violet to

red, that appears when white light passes through a diffraction grating.

stamen
the male reproductive structure of a flower. It consists of the anthers, which produce pollen, and of a slender filament.

stigma
the top part of the carpel where pollen is deposited.

stirrup
the third in a series of three small bones in the middle ear.

style
in plants, the slender, tube-like part of the carpel.

termites
insects similar to ants who live in large colonies. Termites feed on wood and can damage wooden structures.

trend
something that occurs over and over again.

ultraviolet (UV) light
a kind of light not visible to the human eye

valid
well-grounded or justifiable.

vibration
the back-and-forth movement of molecules.

visible spectrum
the part of the electromagnetic spectrum that can be seen by the human eye.

wavelength
the size of a wave, measured from one compression to another compression, or from one rarefaction to another rarefaction.

working hypothesis [plural: hypotheses]
an initial guess about why things work in a certain way. Working hypotheses need to be tested with experiments to see if they are correct.

Glosario

adaptación (adaptation)
una característica especial que le permite a un animal sobrevivir en su medioambiente.

amplitud (amplitude)
la propiedad de una onda de sonido relacionada a cuán unidas están las moléculas.

androceo (androecium)
véase estambre.

angiosperma (angiosperm)
uno de los grupos principales de plantas florales modernas. Éstas producen semillas en órganos reproductivos especializados llamados flores.

anteras (anthers)
las partes en el estambre de una flor donde se produce el polen.

bastoncillo (rod)
un tipo de célula encontrada en la retina. Cada bastoncillo es sensitivo a niveles bajos de luz, pero no puede ver colores.

biólogo (biologist)
el científico que estudia los seres vivientes.

caloría (calorie)
la cantidad de energía necesaria para aumentar la temperatura de un gramo de agua por 1°C.

Caloría (Calorie)
la cantidad de energía en los alimentos. Una Caloría es lo mismo a 1 kilocaloría ó 1000 calorías.

carga (load)
el objeto siendo movido, o que resiste el movimiento de la palanca.

carnívoro (carnivore)
un animal que solamente come carne.

categoría (category)
un conjunto o clase de objetos con características, propiedades o atributos similares.

células receptoras (receptor cells)
las células que reciben información del mundo y la envían al cerebro.

cilios (cilia)
filamentos minúsculos que cubren la cóclea y ayudan a convertir las vibraciones en impulsos nerviosos.

cóclea (cochlea)
un tubo espiral que constituye la parte del oído interno, responsable de la audición.

colaborar (collaborate)
trabajar en conjunto.

colonia (colony)
un grupo de organismos similares viviendo o creciendo juntos.

comportamiento (behavior)
la respuesta de un animal a su
medioambiente.

comportamiento aprendido
(learned behavior)
un comportamiento obtenido
a través de enseñanzas o
experiencias.

comportamiento instintivo
(instinctive behavior)
un comportamiento con el cual los
animales nacen.

compresión (compression)
parte de una onda de sonido donde
las moléculas se compactan.

conducto auditivo (ear canal)
estructura en forma de tubo que
conecta el oído externo al tímpano.

cono (cone)
un tipo de célula que se encuentra
en la retina. Cada cono es sensible
a un color: rojo, verde o azul.

corola (corolla)
la colección de pétalos de color de
una flor.

coyuntura (joint)
el lugar donde dos huesos se unen.

coyuntura movible (movable joint)
coyunturas que se mueven
libremente.

cría (calf)
los jóvenes de algunos tipos de
animales, tales como un elefante.

criterio (criteria)
metas que se deben satisfacer para
lograr exitosamente un reto.

datos (data)
medidas u observaciones
registradas.

decibel (decibel)
unidad utilizada para medir la
sonoridad.

depredador (predator)
un organismo que captura y se
come parcial, o totalmente, otro
organismo.

enzimas (enzymes)
sustancias orgánicas que ocasionan
cambios químicos en otras
sustancias.

especies (species)
todas las plantas o animales de un
tipo específico con características
comunes, tales como los perros.

espectro (spectrum)
una franja de colores, del violeta
al rojo, que aparece cuando la
luz blanca pasa a través de una
retícula de difracción.

espectro electromagnético
(electromagnetic spectrum)
el alcance de las longitudes de
onda emitidas por el sol.

espectro visible (visible spectrum)
la parte de un espectro
electromagnético que puede ser
vista por el ojo humano.

estambre (stamen)
la estructura reproductiva
masculina de una flor. Consiste de
las anteras, las cuales producen
polen, y de un filamento delgado.

estigma (stigma)
la parte superior del carpelo donde se deposita el polen.

estilo (style)
la parte delgada, en forma de tubo, del carpelo.

estribo (stirrup)
el tercero de una serie de tres huesos pequeños en el oído medio.

etnógrafo (ethnographer)
un científico que estudia a las personas.

etograma (ethogram)
una tabla utilizada para registrar las observaciones del comportamiento animal.

etologista (ethologist)
un biólogo que estudia el comportamiento de los animales en su ambiente natural.

explicación (explanation)
una afirmación que conecta una reclamación con la evidencia y el conocimiento científico.

fertilizar (fertilize)
en biología, unir las células sexuales masculinas y femeninas para formar un organismo nuevo.

forraje (forage)
la búsqueda de algo, especialmente alimentos y provisiones.

forrajeador (forager)
un organismo que busca alimento.

frecuencia (de ondas)
(frequency [of waves]) el número de ondas que pasan por un punto en un segundo.

fuerza de esfuerzo (effort force)
la fuerza aplicada a una palanca.

glándulas mamarias
(mammary glands)
glándulas productoras de leche encontradas en las hembras de los mamíferos y utilizadas para alimentar a las crías.

herbívoro (herbivore)
un organismo que se alimenta solamente de plantas.

hertcio (Hz) (hertz [Hz])
una unidad utilizada para medir frecuencias. Para la longitud de onda, es igual a una onda por segundo.

hipótesis de trabajo (working hypothesis [plural: hypotheses])
una conjetura acerca del por qué las cosas trabajan de cierta manera. Las hipótesis de trabajo necesitan ser demostradas con experimentos para ver si son correctas.

homeostasis (homeostasis)
el mantenimiento de las condiciones internas estables en un organismo.

interpretación (interpretation)
una explicación o establecimiento del sentido o significado de algo.

iteración (iteration)
una repetición que intenta mejorar un proceso o producto.

larvas (larvae [singular, larva])
aquellos insectos acabados de nacer (en este caso, la abeja).

limitaciones (constraints)
factores que limitan cómo tú puedes realizar el reto.

longitud de onda (wavelength)
medida de una compresión a otra, o de una rarefacción a otra.

luz ultravioleta (UV) (ultraviolet [UV] light)
un tipo de luz que no es visible al ojo humano.

mamífero (mammal)
un animal de sangre caliente, con pelo, y cuya hembra tiene glándulas especiales para alimentar sus crías con leche.

mamífero marino (marine mammal)
un mamífero que vive en el mar y/u obtiene su alimento del mar.

manada (pod)
un grupo social de ballenas o delfines. Los miembros de una manada pueden protegerse los unos a los otros.

máquina (machine)
un aparato que hace el trabajo más fácil.

martillo (hammer)
el primero de una serie de tres huesos pequeños en el oído medio.

membrana (membrane)
una capa de tejido que sirve como cubierta, conexión o forro.

metabolismo (metabolism)
la combinación de reacciones químicas a través de las cuales un organismo desarrolla o rompe los materiales convirtiéndolos en energía para llevar a cabo sus procesos de vida.

modelo (model)
una representación de algo en el mundo.

mutualismo (mutualism)
una relación entre organismos de dos especies diferentes en la cuál cada miembro se beneficia.

néctar (nectar)
un líquido azucarado producido por las plantas.

nectario (nectary)
el órgano de las plantas que segrega el néctar. El nectario normalmente se encuentra dentro de la flor.

observar (observe)
usar uno de los cinco sentidos para recoger información sobre un objeto o fenómeno.

oceanógrafo (oceanographer)
un científico que estudia el océano.

oído (ear) el órgano sensorial encargado de recibir y procesar ondas de sonidos en impulsos nerviosos.

ommatidio (ommatidia)
uno de los elementos estructurales que constituyen el ojo compuesto de los insectos.

omnívoro (omnivore)
un organismo que se alimenta de diferentes clases de alimentos, incluyendo plantas y animales.

onda sonora (sound wave)
el movimiento de moléculas en un patrón, repetido una y otra vez muy rápidamente.

ovario (ovary)
la parte agrandada en la base del carpelo que contiene óvulos o huevos.

óvulo (ovule)
una estructura diminuta como un huevo en las plantas florales que se desarrolla en una semilla después de la polinización.

palanca (lever)
una máquina simple hecha de una barra rígida y fulcro, o un punto de equilibrio.

perchas (roosts)
lugares comunales de descanso, mayormente para una sola especie de pájaros.

pichón (hatchling)
un pajarito muy joven.

pistilo de carpelo (carpel of pistil)
el órgano reproductivo femenino de una flor. Se puede componer de un carpelo sencillo, de dos o más carpelos unidos.

polen (pollen)
granos pequeños y polvorientos que contienen las células sexuales masculinas en las plantas con semillas.

polinización (pollination)
la transferencia de polen (células sexuales masculinas) de un estambre al estigma de una flor.

polinizador (pollinator)
un insecto u otro animal que transporta polen de una planta a otra.

probóscide (proboscis)
órgano de alimentación y de succión, delgado y tubular de ciertos insectos, tales como las mariposas y la polilla.

radiación electromagnética (electromagnetic radiation)
una onda que viaja a través del espacio y lleva energía.

rarefacción (rarefaction)
parte de una onda sonora donde las moléculas se esparcen.

receptáculo (receptacle)
la base de una flor a la cuál las partes de la flor están adheridas.

recomendación (recommendation)
una aseveración que sugiere qué hacer en una situación descrita.

regurgitar (regurgitate)
llevar comida parcialmente digerida del estómago a la boca.

reproducir (reproduce)
producir descendencia.

retina (retina)
una membrana sensitiva a la luz
que cubre la parte de atrás del
interior del ojo.

sabana (savanna)
una ciénaga con árboles dispersos,
generalmente encontrados
en regiones tropicales o
subtropicales.

salto (breaching)
un comportamiento observado
en mamíferos marinos y algunos
peces, donde el animal salta alto
sobre la superficie del agua y cae
salpicando mucha agua.

sépalos (sepals)
partes verdes parecidas a una hoja
de una flor encontrada fuera de
los pétalos.

simulación (simulation)
uso de un modelo para imitar o
actuar situaciones de la vida real.

sonar (sonar)
una técnica que usa sonido para
proveer imágenes de objetos que
se encuentran debajo del agua.

**sonoridad (o intensidad de
sonido)** (loudness [or intensity])
cuán alto o bajo es un sonido.

tendencia (trend) **algo (en este
caso comportamiento)**
que ocurre una y otra vez.

termitas o comején (termites)
insectos similares a la hormiga,
que viven en grandes colonias.
Las termitas se alimentan
de madera y pueden dañar
estructuras de madera.

tímpano (eardrum)
la membrana que separa el oído
exterior del oído interior.

tono (pitch)
una medida de la frecuencia de la
vibración de la procedencia de un
sonido.

tubo de polen (pollen tube)
el tubo delgado formado por los
granos de polen para alcanzar y
fertilizar los óvulos.

válido (valid)
bien fundado o justificable.

varado o encallado (breached,
breaching, to breach)
cuando un mamífero marino que
no puede vivir fuera del agua se
encalla en tierra, usualmente una
playa.

ventana vestibular (oval window)
una abertura entre el oído medio y
el oído interno.

vibración (vibration)
el movimiento de un lado a otro de
las moléculas.

yunque (anvil)
el segundo en una serie de tres
huesos pequeños en el oído medio.

Index

A

adaptation
defined, AIA 32
of predators, AIA 86-AIA 90

amplitude, defined, AIA 127

angiosperms, defined, AIA 78

anthers, AIA 74-AIA 75
defined, AIA 79

anvil (bone in ear),
defined, AIA 130

B

ball-and-socket joints,
AIA 88-AIA 89

beaching, defined, AIA 139

bees
communication of,
AIA 109-AIA 113
foraging of
life in hive, AIA 62-AIA 63
mutualism with flowering
plants, AIA 74-AIA 77
vision in bees, AIA 69

behavior
See also communication;
feeding; survival
defined, AIA 32

biologists, defined, AIA 3

birds
raising young, AIA 34-AIA 35
roosts, AIA 33

body structure
and communication, AIA 114
dolphins, AIA 133-AIA 134,
AIA 135
elephants, AIA 120-AIA 123
and feeding, AIA 44-AIA 45,
AIA 46, AIA 56, AIA 58
bees, AIA 61, AIA 62-AIA 63,
AIA 71, AIA 110
carnivores, AIA 82-AIA 83,
AIA 86-AIA 90
chimpanzees, AIA 45,
AIA 47, AIA 54, AIA 56,
AIA 60

breaching, defined, AIA 138

C

calorie, defined, AIA 33

Calorie, defined, AIA 33

calves, defined, AIA 121

carnivores
defined, AIA 44
feeding behaviors of,
AIA 82-AIA 90

carpel, defined, AIA 79

categories, AIA 16-AIA 17
defined, AIA 16

cheetahs, feeding behaviors of,
AIA 82, AIA 84-AIA 85, AIA 90

chimpanzees, feeding behaviors,
AIA 44, AIA 45, AIA 47-AIA 60

cilia, defined, AIA 130

claims, AIA 28-AIA 29
See also explanations

cochlea, defined, AIA 130

collaborate, defined, AIA 55

collecting data *See* data collection

colonies, defined, AIA 62

color, and light, AIA 72-AIA 73

Project-Based Inquiry Science

communication, AIA 97-AIA 101
  bees, AIA 109-AIA 113
  elephants, AIA 118, AIA 119,
  AIA 120-AIA 123
  forms of animal, AIA 114,
  AIA 119
  forms of human, AIA 102-AIA 108
  marine mammals,
  AIA 131-AIA 135,
  AIA 136-AIA 141
  as survival behavior, AIA 98
compression, AIA 126-AIA 127
  defined, AIA 124
cones, defined, AIA 73
conservation, AIA 4
  defined, AIA 5
constraints
  defined, AIA 7
corolla, defined, AIA 79
criteria, defined, AIA 7
crocodiles, feeding behaviors of,
  AIA 82-AIA 86, AIA 90
D
Data, defined, AIA 12, AIA 18
  data collection, AIA 12
planning, AIA 15-AIA 19
David Greybeard (chimpanzee),
  AIA 54 AIA 55, AIA 56
debate, scientific
  See waggle dance, of bees
decibels, defined, AIA 128
decision-making,
  by bees, AIA 62 AIA 63
design, of enclosure
  See Addressing the Big Challenge
dolphins, communication of,
  AIA 131-AIA 135,
  AIA 136-AIA 141

E
ear canals, defined, 129
eardrums, defined, 129
ears, in humans, AIA 129-AIA 130
defined, 129
echolocation, AIA 138-AIA 139,
  AIA 141
  defined, AIA 137
effort force, defined, AIA 87
electromagnetic radiation,
  defined, AIA 72
electromagnetic spectrum,
  AIA 72-AIA 73
  defined, AIA 72
elephants, communication of,
  AIA 114-AIA 119,
  AIA 120-AIA 123
ellipsoid joints, AIA 89
embryo, AIA 74
enclosures, AIA 4-AIA 6
  criteria and constraints for,
  AIA 93, AIA 146-AIA 147,
  AIA 149-AIA 150
  defined, AIA 5
environment
  See also enclosures
  effects on communication,
  AIA 131-AIA 132,
  AIA 136-AIA 137, AIA 140
enzymes, defined, AIA 35
ethnographers, defined, AIA 105
ethogram, defined, AIA 15
ethologists, defined, AIA 12
eyes
  of bees, AIA 69
  of humans, AIA 73

F

feeding
  See also specific animal;
  What Affects How Animals Feed
  as survival behavior, AIA 33
  ways of, AIA 44, AIA 46

fertilization
  defined, AIA 74
  in flowering plants, AIA 79

field observations, AIA 53-AIA 55

flowers
  See also bees, foraging of
  anatomy of, AIA 79
  Flower Dissection, AIA 78-AIA 81

foragers
  chimpanzees as, AIA 56
  defined, AIA 44

foraging
  See also bees, foraging of
  defined, AIA 44

frequency (waves),
  AIA 126-AIA 127, AIA 129
  defined, AIA 126

fulcrum, defined, AIA 87

G

gliding joints, AIA 89

Gombe (Africa), AIA 53-AIA 54,
  AIA 55

Goodall, Jane, AIA 53-AIA 60

H

habitat
  See also environment
  defined, AIA 5
  difficulties in observation, AIA 52
  as factor in feeding, AIA 56,
  AIA 58

hammer (bone in ear), defined,
  AIA 130

hatchlings, defined, AIA 35

hearing
  human, AIA 129-AIA 130
  marine mammals, AIA 137

herbivores, AIA 82
  adaptations for feeding, AIA 89
  defined, AIA 44

hertz (Hz), defined, AIA 127

hinge joints, AIA 89, AIA 90

homeostasis, defined, AIA 34

humans
  communication, AIA 102-AIA 108
  comparison with animal behavior,
  AIA 20-AIA 27
  hearing in, AIA 129-AIA 130
  vision in, AIA 72-AIA 73

I

instinctive behavior
  defined, AIA 32
  and raising young, AIA 35

intensity, defined, AIA 128

interpretation
  *See also specific observations
  and explanations*
  defined, AIA 22
  observations and, AIA 22-AIA 23
  vs. observation, AIA 119, AIA 155

Investigation Expos,
  AIA 50-AIA 51
  defined, AIA 50

iteration, defined, AIA 38

J

joints, AIA 86-AIA 90
  defined, AIA 87

L

larvae, defined, AIA 62

Leakey, Louis, AIA 53

learned behavior, defined, AIA 32

levers, AIA 87-AIA 90
  defined, AIA 87

light
  See also electromagnetic spectrum
  and bee vision, AIA 69

lions, feeding behaviors of, AIA 82,
  AIA 83-AIA 86, AIA 90

load, defined, AIA 87

loudness, AIA 127-AIA 128
  defined, AIA 128

M

machines, AIA 86-AIA 90
  defined, AIA 86

mammals
  See also specific animal
  defined, AIA 32
  reproduction and, AIA 34

mammary glands, defined, AIA 34

marine mammals
  communication of,
  AIA 131-AIA 135,
  AIA 136-AIA 141
  defined, AIA 131

membranes, defined, 129

metabolism, defined, AIA 33

middle ear, AIA 130

models, AIA 61-AIA 62
  bee foraging, AIA 64-AIA 67,
  AIA 68
  defined, AIA 61

movable joints
  defined, AIA 87
  types of, AIA 88-AIA 90

mutualism, AIA 75, AIA 77
  defined, AIA 75

N

nectar, AIA 62-AIA 63,
  AIA 64-AIA 67
  defined, AIA 61

non-verbal communication,
  AIA 102-AIA 108, AIA 114

O

observe, defined, AIA 3

oceanographers, defined, AIA 138

odors
  See also scent markers
  as communication, AIA 114,
  AIA 121

omnivores, AIA 82
  See also specific animal
  adaptations for feeding, AIA 89
  defined, AIA 44

oval window, defined, AIA 130

ovary, AIA 74
  defined, AIA 79

ovules, AIA 74
  defined, AIA 79

P

pistil, AIA 74
  defined, AIA 79

pitch, AIA 126, AIA 128
  defined, AIA 126

pivot joints, AIA 89

pods, defined, AIA 131

pollen, AIA 62-AIA 63,
  AIA 64-AIA 67
  allergic reactions and, AIA 76
  defined, AIA 61

pollen tube, defined, AIA 79

pollination, AIA 74-AIA 75
 defined, AIA 74
 wind pollination, AIA 76
pollinators, AIA 75
predators
 adaptations for feeding,
 AIA 86-AIA 90
 defined, AIA 44
 feeding behaviors of,
 AIA 82-AIA 90
Project Board, defined, AIA 8
protection, survival behavior,
 AIA 33-AIA 34

R

rarefaction, AIA 126-AIA 127
 defined, AIA 124
receptacle, defined, AIA 79
receptor cells, defined, AIA 69
recommendations, defined,
 AIA 91
regurgitation, defined, AIA 35
reliable data, defined, AIA 18
reproduction
 defined, AIA 74
 in flowering plants,
 AIA 74-AIA 76, AIA 79
 survival behavior, AIA 34-AIA 35
retinas, defined, AIA 73
rods, defined, AIA 73
roosts, defined, AIA 33
rumbles, elephant sounds,
 AIA 120-AIA 121

S

saddle joints, AIA 89
safety *See* protection
savanna, defined, AIA 115
scent markers, of bees,
 AIA 111-AIA 112, AIA 113

science knowledge,
 AIA 28-AIA 29
 See also explanations
seeds, AIA 74
sepals, defined, AIA 79
signature whistles, of dolphins,
 AIA 137
simulations, AIA 61-AIA 62
 bee foraging, AIA 64-AIA 67,
 AIA 68
 defined, AIA 61
smell *See* odors
**sonar (SOund NAvigation and
Ranging)**, AIA 137,
 AIA 138-AIA 139, AIA 141
 defined, AIA 138
sound
 dolphin communication,
 AIA 136-AIA 141
 elephant communication,
 AIA 120-AIA 121, AIA 123
 and human hearing,
 AIA 129-AIA 130
 wave nature of, AIA 124-AIA 128
sound waves
 See also echolocation; sonar
 defined, AIA 124
species, defined, AIA 32
stamen, defined, AIA 79
stigma, AIA 74-AIA 75
 defined, AIA 79
stirrup (bone in ear), defined,
 AIA 130
style, AIA 74
 defined, AIA 79
survival, animal needs for,
 AIA 32-AIA 37, AIA 98

T

tables, in recordkeeping, AIA 25

termites
and chimpanzee feeding, AIA 54,
AIA 56-AIA 57, AIA 60
defined, AIA 54

tool use, by chimpanzees, AIA 56

touch, as communication,
AIA 121, AIA 138

trends, AIA 16-AIA 17
defined, AIA 16

U

ultraviolet (UV) light,
AIA 72-AIA 73
defined, AIA 69

V

validity, defined, AIA 30

verbal communication,
AIA 102-AIA 108, AIA 114

vibrations
defined, AIA 124
and hearing, AIA 130

vision
in bees, AIA 69
in humans, AIA 72-AIA 73

von Frisch, Karl, on bee waggle
dance, AIA 109-AIA 112

W

waggle dance, of bees,
AIA 109-AIA 113

wavelengths, AIA 126-AIA 127
defined, AIA 126

Wenner, Adrian, on bee waggle
dance, AIA 111-AIA 112

wind pollination, AIA 76

Y

young, raising of, AIA 34-AIA 35

Z

zoos
changes in, AIA 4-AIA 6

84 Business Park Drive, Armonk, NY 10504
Phone (914) 273-2233 Fax (914) 273-2227
www.its-about-time.com

Staff Credits

President
Tom Laster

Director of Product Development
Barbara Zahm, Ph.D.

Managing Editor
Maureen Grassi

Project Development Editor
Ruta Demery

Project Manager
Sarah V. Gruber

Development Editor
Francesca Casella

Assistant Editors
Gail Foreman
Susan Gibian
Rhonda Gordon
Nomi Schwartz

Safety and Content Reviewer
Edward Robeck

Creative Director
John Nordland

Production/Studio Manager
Robert Schwalb

Layout and Production
Tommaso Gaetani

Illustration
Dennis Falcon

**Technical Art/
Photo Research**
Sean Campbell
Michael Hortens
Marie Killoran

Equipment Kit Developer
Dana Turner

Picture Credits

Page 4   Wildlife Conservation Society

Page 5   Smithsonian Institution

Page 6   Janis Clark/PAWS:
            www.paws.org

Page 14   Fotolia/Impala

Page 19   Fotolia/Mazur Serghei

Page 21   Big Stock/Siok Hian

Page 44   It's About Time

Page 49   Jane Goodall Institute

Page 53   Fotolia/Mark Hughes

Page 55   Jane Goodall Institute

Page 62   National Aeronautics
            and Space A

Page 69   Nakashima Eye Center/
            Yasuyuki Nakashima

Page 75   istockphoto/arlindo71

Page 76   istockphoto/Christopher Steer

Page 76   Laresen Twins Orchids, DK

Page 83   Fotolia/Irene Teesalu

Page 84   Fotolia/Jean François Le Fevre

Page 95   abc News Productions

Pages   1, 3, 5, 6, 11, 17, 21, 24, 26,
            32, 33, 34, 35, 43, 44, 47, 52,
            56, 61, 63, 69, 75, 76, 77, 78,
            79, 85, 88, 89, 90, 91, 92, 94,
            101, 102, 106, 112, 113, 115,
            118, 119, 124, 125, 126, 131,
            136, 137, 141, 142, 144
            istockphoto

All Illustrations Dennis Falcon

Technical Art   Marie Killoran/
                 Michael Hortens/
                 Sean Campbell